OPEN TUBULAR COLUMN
GAS CHROMATOGRAPHY IN
ENVIRONMENTAL SCIENCES

OPEN TUBULAR COLUMN GAS CHROMATOGRAPHY IN ENVIRONMENTAL SCIENCES

Francis I. Onuska

National Water Research Institute
Canada Centre for Inland Water
Burlington, Ontario, Canada

and

Francis W. Karasek

University of Waterloo
Waterloo, Ontario, Canada

PLENUM PRESS • NEW YORK AND LONDON

Library of Congress Cataloging in Publication Data

Onuska, Francis I., 1935–
  Open tubular column gas chromatography in environmental sciences.

  Includes bibliographical references and index.
  1. Gas chromatography. 2. Pollutants — Analysis. I. Karasek, Francis W., 1919–   . II.
Title.
QD79.C45O58   1984   543′.0896   84-4806

ISBN-13: 978-1-4684-4690-6   e-ISBN-13: 978-1-4684-4688-3
DOI: 10.1007/978-1-4684-4688-3

©1984 Plenum Press, New York

Softcover reprint of the hardcover 1st edition 1984

A Division of Plenum Publishing Corporation
233 Spring Street, New York, N.Y. 10013

PREFACE

The method of choice for analysis of organic pollutants in the environment is gas chromatography, using one of the most common detection systems such as flame ionization, electron capture, and mass spectrometric detection.

Perhaps the most prevalent difference in practice among the environmental analytical chemists involves the type of column used in gas chromatography. The traditional packed-column gas chromatography is rapidly being replaced with wall-coated open tubular (WCOT) columns. However, in using WCOT columns there are difficulties in reporting analytical data and the means of quantitation.

This book is intended to present a survey of the salient features of WCOT column gas chromatography that relate to the selection of WCOT columns, their technology, separation efficiency, and determination of the principal groups of organic micropollutants, such as hydrocarbons, aromatics, polynuclear aromatic hydrocarbons, organochlorine pollutants, chlorinated phenols, dioxins, dibenzofurans, ureas, carbamates, and odor-causing substances in complex environmental matrices. Hence, it is the hope of the authors that the content of this book will be of significance to a wide spectrum of disciplines that include environmental chemistry, industrial hygiene and health, toxicology, genetics, and industry.

The major thrust of the book is directed to the gas chromatographer who must continuously bear the major burden of identification and quantitation of an enormous number of environmental contaminants. Undoubtedly, WCOT column gas chromatography will continue to be the most reliable and useful separation technique at his disposal. However, we have also endeavored to produce a reasonably complete text for all the steps of analysis by WCOT column technology for environmental samples by adding briefly the basic principles of cleanup techniques, preconcentration, and guidelines for qualitative analysis. In addition, general methods which have increased in importance within the last few years, such as the use of computers and interfacing with mass spectrometry, are also covered. Throughout we have concentrated on the principles involved.

v

The material presented in the various chapters is supported by extensive references. It is hoped that these references will enable readers to obtain an in-depth treatment of those aspects of the analytical methods that may be important for solution of their specific problems.

In summary, it has been our aim to provide a textbook which could furnish the necessary background for all readers interested in high-resolution gas chromatography, whether from the viewpoint of an analytical method or of the separation process involved, or both. The critical insights and up-to-date information should be of value to those already working in the field.

FRANCIS I. ONUSKA
FRANCIS W. KARASEK

ACKNOWLEDGMENTS

Our thanks are due to the following journals, companies, and individual scientists for their permission to use diagrams and tables from their research works and scientific papers:

Analytical Chemistry
Ann Arbor Science Publishers Inc.
Bulletin of Environmental Science and Technology
Chromatographia
Fresenius' Zeitschrift für Analytische Chemie
Journal of Chromatography
Journal of High Resolution Chromatography and Chromatography
   Communications
Carlo Erba Strumentazione
Dani S.p.A.
Finnigan-MAT Instruments Co.
Hewlett-Packard Company
Varian Instrument Division
Prof. Kurt Grob, EAWAG, Dübendorf, Switzerland
Prof. Victor Protorius, University of Pretoria, South Africa
Prof. Milton L. Lee, Brigham Young University, Provo, Utah, U.S.A.
Dr. David L. Stalling, Columbia National Fisheries Research
        Laboratory, Columbia, Missouri, U.S.A.

CONTENTS

CHAPTER SIX: THE RETENTION INDEX SYSTEM

CHAPTER SEVEN: QUANTITATIVE ANALYSIS

CHAPTER EIGHT: APPLICATION IN ENVIRONMENTAL ANALYSIS

INTRODUCTION

1.1. ORGANIC CONTAMINANTS IN THE ENVIRONMENT

The many benefits of our modern, industrial society are accompanied by some political, economical, and environmental hazards. Careful assessment of the relative risk of existing environmental hazards is necessary for the establishment of regulatory policies. These regulations serve to enhance the quality of our environment. Thus, one of their primary objectives is to establish and maintain a broad surveillance and evaluation program concerned with the extent and significance of the contamination of man and his environment by organic and inorganic compounds.

The analysis of organic contaminants in our environment involves the determination of the amount and type of carbon-containing molecules in the ecosystem. In a broad sense an ecosystem is defined as an enclosed system in which all needs are provided from within the system. The science that deals with this aspect of the environment and its subcompartments, water, air, and aquatic sediments is called ecology.

Organic contaminant sources in the environment could be classified into four categories[1]:

1. All organics from the atmosphere,
2. Man-made organic materials,
3. By-products of chlorination, and
4. Biological metabolites and breakdown products.

Environmental chemists call these compounds xenobiotics. A xenobiotic compound may be defined as a compound, natural or man-made, that has been added by man to some subcompartment of the biosphere. A prime responsibility of environmental analytical chemists is to create new and improved analytical procedures suitable for the monitoring of low concentrations of organic contaminants. It is important that uniform chemical methodology of high reproducibility and accuracy be used to ensure that analytical results can be correlated and compared between laboratories.

The ethical aspects of hazardous organic contaminants are magnified because protection of human beings and the environment must be balanced against a cost limitation. The expansion of ethical aspects to the

1

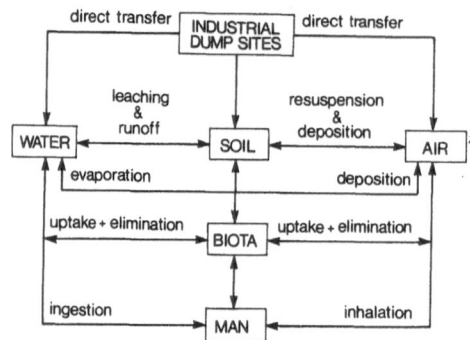

FIGURE 1.1. Environmental pathways of hazardous organic contaminants.

biosphere is studied by bioethics. As a discipline, it addresses the new set of problems arising from the human struggles to enhance and control the quality of life and well-being. Careful and detailed risk analyses can be made from results obtained by environmental analytical chemists, toxicologists, and physicians. From these data it is necessary to set priorities to those areas where real risks exist. Analytical chemists must necessarily separate scientific facts and analytical data from speculation, opinion, and personal or employer bias.

A less obvious but perhaps equally insidious danger confronts the analytical chemist whose data interpretation is subjected to administrative review prior to publication of data. The analytical environmental chemist has a responsibility to ensure that in the interests of a balanced view, the significance of any specific analytical study does not become obscured.

Trace contaminants can be taken up rapidly by components of the aquatic system, with the largest reservoirs usually in sediments. Major biological and physical transport pathways of hazardous organic chemical contaminants are shown in Fig. 1.1. Examination of the interlinking of these pathways reveals the complexity of environmental studies.

1.2. ADVANCES OF TRACE ORGANIC ANALYSIS

Trace organic analysis has made tremendous advances in the last 10 years but is still far from being able to detect all organics present in the environment. The minimum detection limit of many current analytical techniques is about one microgram per liter or kilogram. It might be assumed that for each order of magnitude that analytical sensitivity is lowered, an additional order of magnitude of organic contaminants will be detectable in environmental samples. By the time a sensitivity of one

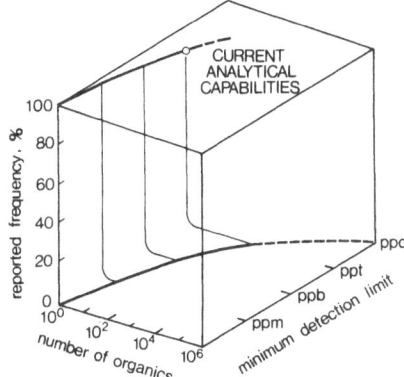

FIGURE 1.2. Schematic representation of current analytical capabilities in organic trace analyses as a function of detection limits.

picogram per liter or kilogram is reached, if the number of known organic chemicals is approximately 10^7, a significant number will be detectable in the environment at least some of the time. At this point the tip of the solid line region will reach the ultimate point, indicated by a circle in Fig. 1.2. As analytical methodology advances even further, that point will become part of the solid area and ideally all known organic contaminants can be identified.

It must be clear that only a tip of the problem has been uncovered with regard to organic contamination in the environment. However, a considerable amount of information is available on the toxicological and chemical characterization of many of the organic contaminants that have been detected so far, and it is useful to employ that information to classify them accordingly. The criteria used for categorization are polarity, volatility, and molecular weight.

Polarity of an organic molecule reflects the degree to which the molecule possesses a dipole moment. Molecules having acidic or basic functional groups, e.g., —OH, —COOH, —NH_2, usually exhibit strong dipole moments.

Volatility is the physical property of components present in a mixture, which is measured as a partial pressure (P_a) of a component over its solution at a certain molar concentration (x_a) as expressed by Raoult's law:

$$P_a = P_a^0 x_a \tag{1.1}$$

The volatile, polar, and water-soluble organic components are generally not isolated at all from polar matrices such as water but are directly introduced into an analyzer. Less polar volatile organic components are isolated by allowing them to partition into an immiscible, less polar solvent or into the gaseous phase (headspace and stripping techniques) above the polar solvent. After removal of organic contaminants from various matrices, the extract generally contains a complex mixture of components in an organic solvent.

1.2.1. Separation Methods

The most common technique used to separate complex mixtures of organic contaminants is chromatography. Chromatography is a physical method of separation in which the components to be separated are distributed between two phases: one of the phases being a stationary bed of large surface area, the other being a mobile phase of gas or liquid which moves through or along the stationary bed.

Chromatographic separation is usually accomplished by one of four techniques:

1. Adsorption chromatography,
2. Partition chromatography,
3. Size exclusion chromatography,
4. Ion exchange chromatography.

Adsorption is a physical property of matter. It is one of a general class of molecular distribution processes that includes vaporization and the partitioning of a sample between adjacent gas or liquid phases. Adsorption may occur at the surface of either a solid or liquid phase. Adsorption chromatography is best suited to organic compounds containing polar groups that promote selective adsorption. A treatment of the basic principles of adsorption chromatography as related to various separation problems in gas–solid and liquid–solid chromatography has been described.[4,13,14]

Partition chromatography, which is the main subject of this book, is similar to adsorption chromatography. However, the solid phase is coated with a liquid phase, into which the organic component dissolves rather than adsorbs. Of the two gas chromatographic methods, gas–liquid chromatography is the most significant. Gas–solid chromatography is used mainly for the separation and analysis of low boiling point compounds, particularly inorganic gases. Since the instrumental requirements for both methods are practically the same, their practice can be discussed concurrently.

Size exclusion chromatography separates organic molecules by molecular weight through permeation or filtration processes through a column packed with solid particles containing numerous pores. In the permeation process the organic molecules of smaller size spend a larger percentage of time in the pores. Therefore, the organic molecules are selectively eluted from the column, with those of highest molecular weight coming off the column first. In the filtration mode of size exclusion chromatography, the medium contains no pores, and the elution order is reversed. In both modes of operation of size exclusion chromatography, it is important that adsorption effects be avoided because they degrade the size separation.

Ion exchange chromatography involves the interchange of ions in sample with ionic groups on a synthetic organic resin ion exchanger. Those compounds whose ions interact the least with those of the resin elute from the column first. Because ion exchange is an electrostatic attraction process, the compounds suitable for this technique must contain, or be transformed to, compounds containing ionic groups. A summary of the separation techniques used for the different classes of organic compounds is presented in Table 1.1. Partition and adsorption chromatography using gas as the mobile carrier phase are the most common techniques for separating volatile and semivolatile organic compounds. Liquid mobile phases are also used for some of the semivolatile organic compounds in thin-layer chromatography and column chromatography. For the polar organic compounds of high molecular weight, ion exchange chromatography and liquid–solid adsorption chromatography, such as column, high-pressure or thin-layer configurations are often applicable. Larger semipolar organic molecules can be subjected to size exclusion chromatography for separation, whereas this technique is unsuitable for large polar molecules because adsorption degrades the size separation process. Nonpolar, high molecular weight organic compounds are very insoluble and can be separated by size exclusion chromatography in the

TABLE 1.1
Separation of Organic Molecules by Various Chromatographic Techniques

Polarity	Volatile	Semivolatile	Nonvolatile
Polar	Gas–solid adsorption	Gas–liquid, partitioning Liquid–solid, adsorption	Ion exchange Liquid–solid, adsorption
Semipolar	Gas–solid Gas–liquid	Liquid–liquid Gas–liquid, partitioning	Size exclusion, filtration
Nonpolar	Gas–solid Gas–liquid	Gas–liquid, partitioning	Size exclusion, permeation Liquid–liquid
Molecular weight	Low	Medium	High

permeation mode without peak broadening. Liquid–liquid partitioning in the column mode of operation with a column packing coated with a liquid phase is also useful for analysis of some large, nonpolar organic compounds.

Although there is a separation technique possible for each type of organic compound present in the environment, the technique of using gas as the mobile phase is more fully developed than liquid carrier techniques. Gas chromatography dominates the chromatographic separation processes with its superior speed, high resolution, and excellent detector sensitivities.

The recent advances in development of dependable high-pressure pumps which allow the use of microparticle-packed columns has brought about high-resolution separation in liquid column chromatography. This technique is quite suitable for liquid–solid and liquid–liquid chromatographic separation of thermally unstable organic compounds and high molecular weight compounds. In addition, it is capable of separating thermally stable organic compounds suitable for gas chromatography. High-pressure liquid chromatography, or high-performance liquid chromatography (HPLC) has been developing very rapidly. Its recent successful expansion to the area of wall-coated open tubular (WCOT) column separation will be an excellent ancilliary method to reach the full potential of separation capabilities.

1.2.2. Historical Development of WCOT Column Gas Chromatography

In 1956 Martin[2] described the possibility of using very small bore gas chromatography (GC) columns. In 1958 Golay,[3] following his presentation of the mathematical treatment of WCOT columns, showed several chromatograms demonstrating the separation efficiency of an open capillary tube coated with a nonpolar specific stationary phase. This column separated meta- and para-xylenes and gave an efficency of well over 50,000 theoretical plates. The use of capillary tubing has led to the common designation of WCOT columns as capillary columns. Shortly afterward, Dijkstra and DeGoey[5] prepared an 80-m copper capillary column having 0.25-mm i.d. which was coated with squalane and had an efficiency of about 100,000 theoretical plates. This column separated the nine C_7 paraffinic hydrocarbons. This achievement was not surpassed until recently with the success of extensive research to improve the wettability and inertness of the column walls. Successful applications of WCOT columns have solved difficult analytical separations, especially in petrochemical, flavor and environmental chemistry. However, it is quite evident that only a limited amount of work has been reported on the application of WCOT columns in, e.g., pesticide and water analysis. A definite need exists for high-resolution applications in air and water pollution control because of the extraordinary complexity of environmental samples.

The potential of WCOT columns has not yet been utilized to a large extent for environmental and pesticide trace analysis for two fundamental reasons: the difficulty of quantitative injection techniques and an insufficient development of the technology needed to produce inert, thermally stable WCOT columns.

Almost all classes of environmentally important compounds, especially positional isomers, require highly efficient columns and specialized detection methods to effect separation. Otherwise, cumbersome analytical cleanup and fractionation procedures must be employed. These extensive fractionation procedures result in lower accuracy of analytical results due to uncontrolled losses. The situation improved when reliable commercial WCOT columns and column injection systems become available routinely. Although much improvement is still possible in the engineering of direct on-column injection devices, the basic studies have been carried out.[6]

The technology of obtaining optimum surfaces for coating has been advanced successfully by use of gaseous corrosive agents.[7-12] Separation power, film stability, and the inertness of WCOT columns have been significantly improved. The introduction of fused-silica columns has improved surface inertness but has also created different problems in their preparation, such as wettability of the silica surfaces with more polar stationary liquid phases.

Glass WCOT column gas chromatography is presently in an era of rapid growth. The literature today is filled with new applications of this technique for solution of complex problems. A number of review articles are available which describe the development of glass WCOT columns.[13-19]

1.3. NOMENCLATURE AND RELATIONSHIPS IN GAS CHROMATOGRAPHY

Separation is the primary process of a gas chromatograph. This process takes place inside the column. A length of glass capillary in the case of a WCOT column, containing a stationary phase coated on an inner glass surface is spread at a certain film thickness. Separation is achieved by differences in the partition distribution of the components of a sample between the mobile (gaseous) and stationary (liquid) phases, causing them to move through the column at different rates and elute at different retention times.

The chromatographic process is illustrated in Fig. 1.3. The mixture, consisting of two components A and B, is introduced into a gaseous mobile phase, the carrier gas, which is flowing at a constant rate through the length of the capillary column. While moving through the column, components A and B interact with the stationary phase, but not with the mobile phase.

The elution of the components from the column is sensed by means of a suitable detector, and signals are transmitted to a strip chart recorder or an integrating printer–plotter which produces the chromatogram. This chromatogram consists of a series of peaks. Each peak indicates the elution of a component, and its area is generally directly proportional to the amount of component present in a mixture. The time of elution under specified conditions may be used to characterize the components of the mixture. It is defined as the retention time, or elution time, for a particular component. In an ideal situation all of the peaks in a chromatogram should be completely separated. The peak width (w_b) at the base should be as narrow as possible throughout the chromatogram. Performance factors are measured in terms of the separation efficiency, which is given by the number of theoretical plates, n, calculated from the expression

$$n = 16\left(\frac{t_R}{w_B}\right)^2 = 5.54\left(\frac{t_R}{w_{1/2}}\right)^2 \tag{1.2}$$

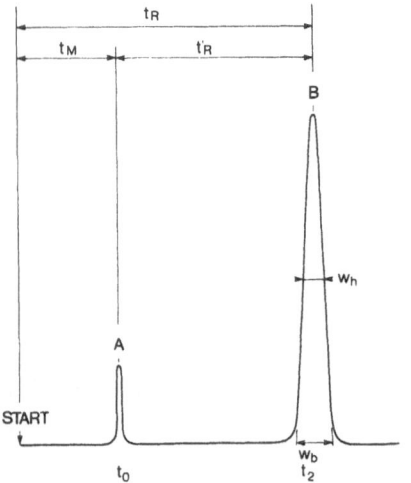

FIGURE 1.3. Basic chromatographic terms.

where t_R is the distance from the point of injection to the peak maximum and w_B is the idealized width of the peak at the baseline or $w_{1/2}$ is the width of the peak at the half height.

The number of plates of the column is a function of column length, among other parameters. To have a value independent of column length, it is commonly expressed in terms of height equivalent to a theoretical plate (HETP). Usually, HETP is given in millimeters and represents the column length (L) divided by the number of theoretical plates (n), as follows:

$$h = \frac{L}{M} \tag{1.3}$$

Efficiency may be expressed in terms of plates per meter, or per foot, where column efficiency is the reciprocal of h.

The effective theoretical plate number, N, indicates column performance based on adjusted retention measurements where $t_R = t_R - t_{air}$. This also takes resolution into account.

$$N = 16(t_R/w_B)^2 = 5.54(t_R/w_{1/2})^2 \tag{1.4}$$

Resolution R_s and separation factor α are defined by Eqs. (1.5) and (1.6). Height equivalent to an effective theoretical plate, H, represents the column length divided by the effective theoretical plate number, expressed in millimeters, when measured at the van Deemter minimum at optimum linear velocity of carrier gas, u_{opt} (see Sec. 1.4.1a).

The ability of a column to separate a pair of two components in a mixture can be measured in terms of resolution R_s, which is a function of peak width and distance between peaks shown in Fig. 1.4. Peak resolution is expressed by the following equation:

$$R_s = 2(t_{R(B)} - t_{R(A)})/(w_{1/2(A)} + w_{1/2(B)}) \tag{1.5}$$

Usually R_s varies with the peaks being measured and is most meaningful for the pair of peaks that are the most difficult to separate.

Separation factor $\alpha_{A/B}$ is defined as the ratio of the distribution coefficients of two components A and B measured under identical conditions:

$$\alpha_{.1/B} = t_{R(B)}'/t_{R(.1)}' = k_{(B)}'/k_{(.1)}' \tag{1.6}$$

where k is defined in Eq. (1.7). By convention, α is always greater than unity.

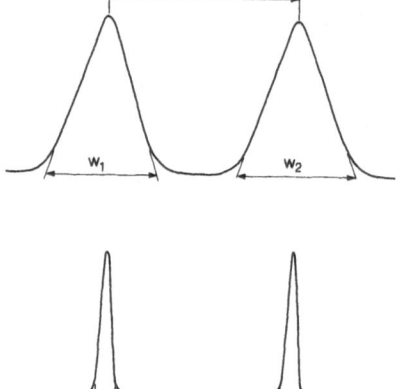

FIGURE 1.4. Resolution between two adjacent peaks.

Partition ratio k is a measure of the distribution of a solute between mobile and stationary phases. It measures the fraction of the sample dissolved in the liquid phase during its travel through the column by the time it spends in the stationary phase (t_R') relative to the time it spends in the gas phase (t_M).

$$k = t_R'/t_M \qquad (1.7)$$

The distribution coefficient K represents the ratio of the concentration of a component in a single definite form in the stationary phase to its concentration in the same form in the mobile phase at equilibrium. Both concentrations are calculated per unit volume of each respective phase at column temperature.

$$K = \frac{\text{amount of solute per gram of stationary phase}}{\text{amount of solute per milliliter of gas}} \qquad (1.8)$$

The ratio of the gas volume to the volume of the stationary phase β is a characteristic physical property of column and is called the phase ratio.

$$\beta = V_M/V_L \qquad (1.9)$$

where V_M is the corrected retention volume of the carrier gas given by the following expression:

$$V_M = V_A \cdot j = t_A \cdot F_c \cdot j \qquad (1.10)$$

where V_L is the volume of the liquid phase in the column, V_A is the retention volume of the inert gas, F_c is the flow rate in ml/min, j is a correction factor (gas compressibility), and p_i/p_0 is carrier gas relative pressure.

$$j = \frac{3}{2} \frac{(p_i/p_0)^2 - 1}{(p_i/p_0)^3 - 1}$$

For practical purposes the phase ratio (β) for open tubular columns can be calculated as follows:

$$\beta = r/2d_f \tag{1.11}$$

where r is the radius of the capillary column and d_f is the stationary phase film thickness. Typical β values for capillary columns are 50–500. For example, a 0.2-mm-i.d. capillary column with a film thickness of 0.2 μm has a β value of 250.

The variation of the number of theoretical plates, n, or effective number of theoretical plates, N, with the partition ratio is illustrated schematically in Fig. 1.5. For closely spaced peaks resolution represents the extent of separation and, according to Purnell,[20] can be written as

$$R = \frac{(\alpha - 1)}{4\alpha} \frac{k'}{(1 + k')} n = \frac{(\alpha - 1)}{4\alpha} \frac{k'}{(1 + k')} \frac{L}{h} \tag{1.12}$$

where k, n, and h refer to the later of two eluting peaks.

Figure 1.6 illustrates the effect of these parameters on four hypothetical chromatograms calculated from Eq. (1.12). A comparison of the

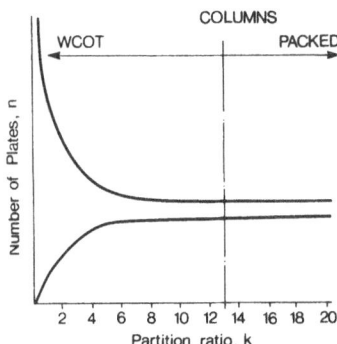

FIGURE 1.5. The relationship of the number of theoretical plates n and the capacity ratio k for a WCOT column.

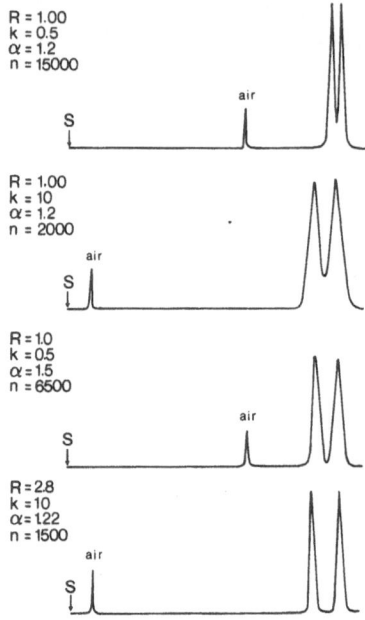

R = 1.00
k = 0.5
α = 1.2
n = 15000

R = 1.00
k = 10
α = 1.2
n = 2000

R = 1.0
k = 0.5
α = 1.5
n = 6500

R = 2.8
k = 10
α = 1.22
n = 1500

FIGURE 1.6. Effect of column efficiency as described by resolution R, the relative retention α, the theoretical plates number n, and the capacity ratio k, in component separation.

first two chromatograms shows that for peaks close to the air (or methane) peak the plate number must be higher than for peaks with large k values in order to obtain the same resolution. Although the partition ratio is the same in the first and third chromatograms, because of the larger relative retention chromatogram, fewer plates are needed in order to obtain the same resolution. Although the relative retention and partition ratio are the same in second and fourth case, the latter shows significantly better resolution due to the higher plate number. The better resolution in the fourth case than in the first, despite the same relative retention and plate number, is due to the higher partition ratio.

, Chromatographic traces in the first and fourth cases are typical for capillary columns. Theoretical treatment dealing with the speed of analysis will be discussed later.

1.4. SEPARATION NUMBER (TRENNZAHL)

Another practical measure of column efficiency which is widely accepted and used for WCOT columns is the separation number (SN) [or in German terminology Trennzahl (TZ)] introduced by R. Kaiser.[21,22]

In practice, this term should express the real efficiency of capillary columns as the number of chromatographic peaks separable in a given range carried out under isocratic conditions. The separation number is defined as the resolution between two adjacent homologs of the n-paraffin series, C_n and C_{n+1}, and is expressed as

$$SN = \frac{t_{R(n+1)} - t_{R(n)}}{w_{(n+1)} + w_{(n)}} - 1 \qquad (1.13)$$

For an accurate calculation of the separation number for the range between $k = 0$ and $k = 10$, the following relationship must be employed:

$$SN_{real} = \frac{\log(w_2/w_1)}{\left[\log\left(\dfrac{n}{5.54}\right)^{1/2} + 1\right] \Big/ \left[\log\left(\dfrac{n}{5.545}\right)^{1/2} - 1\right]} \qquad (1.14)$$

It was shown that in isocratic elution chromatography the separation power decreases exponentially with increasing retention time. Figure 1.7 illustrates that short analysis time with a fast flow and WCOT columns is theoretically the best way to solve analytical separation problems, and that temperature programming is superior to isothermal analysis.

1.4.1. Required Column Parameters

In practical analysis it is often desirable to estimate the required plate number of column length for separation of a given difficulty to separate a component pair. Since $R = 1.5$ represents a complete sepa-

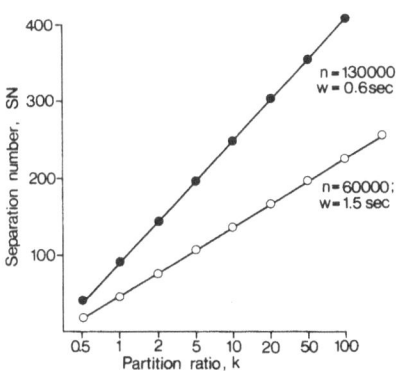

FIGURE 1.7. The relationship between separation number and capacity ratio.

ration, the number of required plates (n_{req}) can be obtained from the equation

$$n_{req} = 16 \cdot R^2[\alpha/(\alpha - 1)]^2 \, [(k' + 1)/k']^2 \qquad (1.15)$$

The number of required plates is often used for comparing the separation of a specific pair of components on various liquid phase coatings. The required length of the column, L_{req}, can be calculated if the plate height is known:

$$L_{req} = 36h[\alpha/(\alpha - 1)]^2[(k' + 1)/k']^2 \qquad (1.16)$$

The practical importance of the mathematical relationship between the height equivalent to a theoretical plate, h, and the linear gas velocity u is expressed by the van Deemter equation[13]:

$$h = A + B/u + C \cdot u \qquad (1.17)$$

where A is the eddy diffusion term describing band broadening caused by the variation of the gas velocity in the porous structure of packed columns. This term in a WCOT column is equal to zero.

The B/u term represents the band broadening as a consequence of longitudinal diffusion of the solute molecules in the gas phase during their residence in the column. The $C \cdot u$ term is related to mass transfer resistance in the column, which hinders the equilibration of the solute molecules between the gas and the stationary phase. The equation is described mathematically by a hyperbola.

For capillary columns the van Deemter equation is expressed as shown in Fig. 1.8. The plot of h (HETP) versus u represents graphically parameters showing band broadening as a mean for interpreting column efficiency in terms of plate height. A detailed mathematical treatment is given by Giddings.[23]

In practice, columns are usually operated above the optimum gas velocity in order to reduce analysis time. Thus, h is dominated by the $C \cdot u$ term and C becomes the most important parameter. The above equations show that in WCOT columns, the C term is directly proportional to the diameter of the capillary as well as to the square of the liquid film thickness and is inversely proportional to the diffusion coefficients of the solute.

Figure 1.9 illustrates a van Deemter plot for a 25-m $\times$ 0.25-mm-i.d. capillary column operated with three carrier gases: nitrogen, hydrogen, and helium. While the lowest value of h (0.18 mm) is obtained with

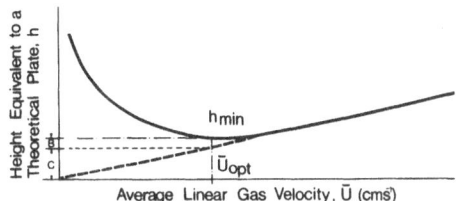

FIGURE 1.8. van Deemter relationship
for a WCOT column.

nitrogen as carrier gas, the speed of analysis is sacrificed since u_{opt} for nitrogen is approximately 9 cm/sec. Using helium as a carrier gas, a small sacrifice in the h value is made ($h = 0.25$ mm), but analysis time is cut in half since the optimum carrier gas velocity is 20 cm/sec. The plot for hydrogen shows that h_{min} is approximately the same as for helium ($h_{min} = 0.26$ mm), but the optimum linear velocity is at 38 cm/sec.

The major advantage of using helium or hydrogen as the carrier gas is that the plate height value rises very slowly as linear velocity of the carrier gas is increased. This means that a much faster analysis with an insignificant loss in resolution may be obtained.

1.4.2. Minimum Analysis Time

A good resolution factor is very costly in terms of analysis time.[24] It takes 3.4 times longer to achieve a resolution of 1.5 than a resolution of unity. In fact, the analysis time is proportional to $N^{3/2}$. An analysis requiring 10^6 plates would be 32,000 times longer than one needing 10^3 plates. Time is the limiting factor in achieving very difficult separations.

The expression for calculating the minimum analysis time is

$$t_R = 9.6R^3(\alpha/\alpha - 1)^3(1 + k')^2(1 + 6k' + 11k'^2)/k'^3 \cdot d_r (2\eta/P_0D_g)^{1/2} \quad (1.18)$$

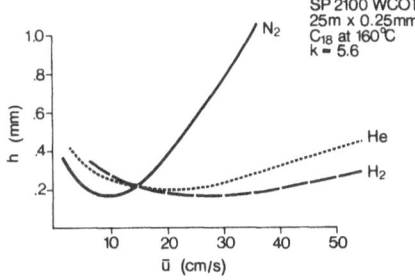

FIGURE 1.9. van Deemter plots for a 25-
m × 0.25-mm-i.d. WCOT column operated with nitrogen, hydrogen, and helium
as carrier gases.

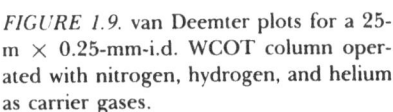

where η is the viscosity of a carrier gas, d_r is the diameter of a capillary column, D_g is the molecular diffusion coefficient for the gaseous phase, and, P_0 = carrier gas absolute pressure at column outlet

The characteristic viscosities of various carrier gases are given in Table 1.2. These values indicate that hydrogen is considerably better than other gases. Hydrogen also has the lowest critical pressure.

Equation (1.18) shows that the analysis time is proportional to the column diameter (d_r) and that the use of smaller bore columns (e.g., 0.065-mm i.d.) permits a fourfold reduction of the analysis time compared to the conventional 0.250-mm-i.d. columns.

1.4.3. Column Length

The minimum column length necessary for a given analysis is determined by the number of plates required for obtaining the desired resolution and by the h value. The latter can be considered constant if conditions and column parameters other than the length are held constant.

Inlet pressure is limited by instrumental factors. Thus, the maximum length of open tubular columns that can be operated at a desired gas velocity is determined by the diameter of the capillary, which in turn governs the permeability of the column. Resolution increases with the square root of the column length. This means that the column length has to be increased four times if the resolution is to be doubled. The following equation shows the relationship between the length of a WCOT column and the inner diameter of the capillary:

$$L = 8.5R^2\left(\frac{\alpha}{\alpha - 1}\right)^2 \frac{1 + k'}{k'^2}\left(\frac{1 + 6k' + 11k'^2}{3}\right) \cdot d_r \qquad (1.19)$$

TABLE 1.2
Viscosities of Various Gases at Different Temperatures (°C)

	Viscosity (μP)			
	20°	100°	200°	300°
Argon	222	271	321	370
Carbon dioxide	147	185	230	270
Helium	196	230	270	307
Hydrogen	88	103	120	140
Nitrogen	175	208	246	281

TABLE 1.3
Relationships Calculated for Capillary Column Parameters for Comparison of Performances[a]

	65 μm		100 μm		250 μm	
α	t_R (sec)	L (m)	t_R (sec)	L (m)	t_R (sec)	L (m)
1.2	$26 \cdot 10^{-3}$	0.07	$3.97 \cdot 10^{-3}$	0.11	0.1	0.27
1.10	$16 \cdot 10^{-2}$	0.23	$24.49 \cdot 10^{-2}$	0.36	0.6	0.90
1.05	1.1	0.85	1.7	1.30	4.3	3.26
1.01	122	19.6	189.6	30.09	473	76.00
1.005	966	78	1494.2	119.18	3750	299.00

[a] Equation employed: $t_R = 0.0184(\alpha/\alpha - 1)^3 \cdot d$, and $L = 29.5(\alpha/\alpha - 1)^2 \cdot d$.

Table 1.3 shows calculated retention times and the calculated corresponding length of WCOT columns capable of separating a pair of components at $k' = 1.76$ at different α. It may be seen from the data in Table 1.3 that small diameter WCOT columns offer greater possibilities for improving their current performances. The optimization of capillary columns as far as the analysis time is concerned is directly proportional to column diameter, while the detection limits are practically constant. It can be seen that the sample size has to decrease in proportion to d_c^2. Small diameter WCOT columns having inner diameters less than 0.1 mm require specially designed inlet systems and detectors, including fast-responding electrometers which give little band broadening so that the separation of very closely related components in the mixture can be achieved.

1.5. THERMODYNAMIC CONCEPT OF POLARITY AND SELECTIVITY

1.5.1. Polarity

The problem of classification of stationary phases and evaluation of their polarity and selectivity has been under scrutiny for over 20 years. Polarities of stationary liquid phases are determined by a high degree of affinity between the liquid phase and a polar compound.[25] The degree of polarity of a stationary phase is determined by its retention of a polar solute relative to a comparable nonpolar solute. The polarity of a compound is associated with its dipole moment. The conclusive factor is defining the polarity of a stationary phase is the extent of orientative (dipole–dipole) interactions. Analysis of the data available in the literature show that the existing definitions of polarity and selectivity are ambiguous and difficult to interpret.

According to the thermodynamic definition, the polarity of stationary phases must be determined from the values of the Gibbs free energies ΔG_1 and ΔG_2 associated with the passage of 1 mole of solute from the stationary phase to the mobile phase.

$$\Delta(\Delta G) = \Delta G_2 - \Delta G_1 = RT \ln \alpha \qquad (1.20)$$

where α is the relative retention, or

$$\Delta(\Delta G) = RT(\alpha - 1) \qquad (1.21)$$

Table 1.4 illustrates how the number of theoretical plates (N) increases with decreasing values of $\alpha - 1$.

It is reasonable to think that changing the nature of the stationary phase will affect ΔG_1 and ΔG_2 in a slightly different way and significantly alter their difference. This is the basic reason why so many different stationary phases are used even with capillary columns.

The polarity classification of Rohrschneider[26] and McReynolds[27] is based on the use of reference compounds of benzene, butanol, 2-pentanone, nitropropane and pyridine, 2-methylpentanol-2, 1-iodobutane, octyne-2, 1,4-dioxan, and *cis*-hydrindane. The polarity of a stationary phase should be evaluated with respect to the parameters of partial molar free energies of solution, ΔG^{CH_2} of the methylene group of an *n*-alkane, and the ΔG values of the five known test substances proposed by Rohrschneider and McReynolds.[28] The test substances given in Table 1.5 simulate the types of intermolecular interactions of functional groups with a stationary phase during the course of analysis.

TABLE 1.4
Calculated Relationships of α and Column Parameters

α	$\Delta(\Delta G)$ (cal/mol)	at $T = 400\ °K$	
		$N\ (k' = 3)$	$N\ (k' = 1.76)$
1.200	146	1,020	1,420
1.100	76	3,440	4,760
1.050	39	12,500	17,530
1.020	16	74,000	102,000
1.015	12	130,000	180,000
1.010	8	290,000	400,000
1.005	4	1,150,000	1,590,000

TABLE 1.5
Type of Interactions for Evaluation of Polarity and Selectivity

Substance	Type of interactions
CH$_2$ group in n-alkane	Dispersive
Benzene	π-complex formation possible
1-butanol	Hydrogen bonding with electron donor groups
2-pentanone	Donor–acceptor complexing
1-nitropropane	Orientative, donor–acceptor complexing
Pyridine	Hydrogen bonding with H donor
	Donor–acceptor complexing

The evaluation of the dispersive polarity values of ΔG^{CH_2} is calculated from the equation

$$\Delta G^{CH_2} = -2.3RT \cdot b \qquad (1.22)$$

where b is equal to log ρ, where ρ is the density of the stationary phase. Values for some stationary phases are shown in Table 1.6. Maximum dispersive polarity, i.e., capacity for dispersive interaction with a substance, is exhibited by such phases as squalane, Kovats C-87 hydrocarbon, and Apiezons.

The capacity of stationary phases for other types of intermolecular interactions including the dispersive interaction is determined from the

TABLE 1.6
Dispersive Polarity of Some Stationary Phases at 120 °C

Stationary phase	b	$-\Delta GCH_2$ (cal/mol)
Squalane	0.2891	520
C$_N$/H$_{1/6}$ hydrocarbon	0.2841	511
Apiezon L	0.2821	507
Apiezon M	0.2833	510
Ucon LB 1715	0.2603	468
Silicone OV-17	0.2551	459
Silicone SE-52	0.2548	458
Silicone SE-30	0.2495	449
Pluronics F 88	0.2362	425
PEG 4000	0.2238	403
Carbowax 20M	0.2235	402
PEG 600	0.2180	392
Carbowax 1000	0.2174	391
DEGA	0.2105	379
Silar 10C	0.1960	353

partial molar free energies of solution of the five test substances listed in Table 1.5. Since no data can be found on the density ρ of a stationary phase at the analysis temperature, values have to be determined as they are essential for an accurate calculation of ΔG. The values of ΔG can be calculated to an accuracy of within 10% if ρ is unknown, employing the following equation

$$-\Delta G + 2.3RT \log \rho = -2.3RT \frac{I - 100n_b}{100} + \log \frac{V_g T}{273} \quad (1.23)$$

where n is number of carbon atoms, V_g is the specific retention volume, and I is the retention index. Table 1.7 indicates polarities of some stationary phases calculated from the ΔG values of five test substances.

There are no pure nonpolar phases. All such phases are capable of inductive interaction with a solute that has a large dipole moment. The higher the value of ΔG for a test substance, the better the capacity of a phase for those intermolecular interactions which are evaluated by the test substance. When polarity is defined in thermodynamic terms, the most polar stationary phase is the one whose ΔG value for each of the five test substances is higher than for all other stationary phases. The most nonpolar stationary phase is the one showing the lowest five values for ΔG of all other stationary phases. It can be seen from Table 1.7 that the most nonpolar stationary phase of those listed is SE-30, while the

TABLE 1.7
Polarity Values of Stationary Phases at 120 °C

Stationary phase	$-\Delta G + 2.3RT \log \rho(cal/mol)$				
	(1)	(2)	(3)	(4)	(5)
Apiezon L	2820	2450	2600	2820	3110
Apiezon M	2820	2560	2610	2810	3100
Dioctyl sebacate	3250	3420	3300	3800	3750
Dioctyl phthalate	3050	3200	3210	3770	3660
Ucon LB 1715	2890	3360	2990	3550	3580
SE-52	2580	2470	2610	2870	2950
SE-30	2530	2410	2540	2740	2850
Pluronics F-88	2950	3530	3020	3880	3810
PEG 4000	2860	3520	2960	3890	3830
Carbowax 20M	2870	3470	2950	3870	3810
Carbowax 1000	2860	3630	3030	3940	3980
DEGA	2460	3080	2680	3550	3700

(1) Benzene, (2) 1-butanol, (3) 2-pentanone, (4) 1-nitropropane, (5) pyridine.

most polar stationary phase is Carbowax 20M. However, each stationary phase has other distinctive characteristics important to WCOT column performances.

1.5.2. Selectivity

Selectivity is a term widely employed in gas chromatography, but its meaning is often not totally understood. The selectivity and efficiency of a column are used interchangeably quite frequently. Sometimes, selectivity is treated as a synonym of polarity[26,27] or interpreted as the capacity to separate two substances with close boiling points. In actuality, the ability of a column to separate substances is a result of both its selectivity and efficiency. Selectivity is related only to the equilibrium characteristics representative of the difference in retention of the analyzed substances and is defined in terms of the relative retention of substances. The selectivity of a stationary phase during separation of two substances is determined as α_{ij}:

$$\alpha_{ij} = \frac{\gamma_j^\infty P_j^0}{\gamma_i^\infty P_i^0} = \frac{e^{\Delta G_i^E/RT}}{e^{\Delta G_i^E/RT} \cdot (P_j^0/P_i^0)} \tag{1.24}$$

where γ_i^∞ and γ_j^∞ are the activity coefficients of substances i and j at infinite dilution, P_i^0 and P_j^0 are the saturated vapor pressures at 1 atm and at the specific column temperature, and ΔG_i^E and ΔG_j^0 are the excess energies of mixing of substances i and j on a given stationary phase.

It is possible to quantitatively evaluate the selectivity of a stationary phase from the difference in excess free energies of solution of the analyzed substances i and j.[28,29] The quantitative evaluation of the selectivity of a stationary phase to a specific test substance is a measure of the specific nature of its interaction with the test substance relative to silicone oil SE-30. The selectivity of some stationary phases with respect to $\delta (\Delta G^E)_{p,SE}$ are shown in Table 1.8. For comparison, the right-hand part of Table 1.8 represents selectivity determined from values of ΔI defined as

$$\Delta I = I_p - I_{squalane} \tag{1.25}$$

where I is the Kovats retention index value:

$$I_i = 100 \frac{\log t_i' - \log t_n'}{\log t_{n+1}' - \log t_n'} + 100n \tag{1.26}$$

TABLE 1.8
Comparison of Selectivities of Selected Stationary Phases

	$(-\delta(\Delta G^E)_{pSE-30} + 2.3RT \log p/p_{SE-30})$					Stationary phase	ΔI_p squalane				
	(1)	(2)	(3)	(4)	(5)		(1)	(2)	(3)	(4)	(5)
	290	40	60	80	260	Apiezon M	31	22	15	30	40
	290	50	70	70	250	Apiezon L	32	22	15	32	42
	360	960	450	810	730	Ucon LB 1715	132	297	180	275	235
	50	60	70	130	100	SE-52	32	72	65	98	67
	10	20	20	20	20	SE-31	16	54	45	65	43
	420	1120	480	1140	960	Pluronics F-88	262	461	306	483	419
	340	1060	410	1130	960	Carbowax 20M	322	536	368	572	510
	330	1220	490	1200	1130	Carbowax 1000	347	607	418	629	589
	-70	670	140	810	850	DEGA	378	603	460	665	658
	-80	590	420	1040	590	Silicone XF 1150	309	620	470	669	528

(1) Benzene, (2) 1-butanol, (3) 2-pentanone, (4) 1-nitropropane, (5) pyridine.

where n (or $n + 1$) is the number of carbon atoms in two n-alkane references. By definition, the Kovats retention index value of an n-alkane is 100 times the carbon number of that n-alkane.

Examination of data in Table 1.8 shows that those stationary phases which have very close selectivity values on most test substances according to McReynolds have quite different selectivities according to the thermodynamic data presented. According to I_p values, silicone XF 1150 and Carbowax 20M are almost identical as far as separation of aromatics is concerned. However, the values of $(\Delta G^E)_{p,SE}$ show significant differences between the two phases, and the specific retention volume is much higher and much more specific and selective to aromatic compounds for Carbowax 20M than for silicone XF 1150.

1.6. STATIONARY PHASES

The most commonly employed liquid phases for capillary columns are the different silicone polymers such as SP-2100, SE-30, OV-101, OV-1, methyl and dimethyl silicone oils, elastomers, and the more polar SE-52, SE-54, OV-17, and Silar types of substituted silicones (Table 1.9).

The use of these liquid phases have the following advantages: (1) the adsorption isotherm is linear so that symmetrical peaks can be obtained, (2) adequate selective phases can be chosen for a specific separation employing criteria discussed earlier, and (3) liquid phases are available in well-defined purity and thus retention data are comparable. The following qualities are important for liquid phases used in WCOT columns: (1) chemical stability at elevated temperatures, (2) sufficient resolving power for the components under study, (3) selectivity, (4) low viscosity at the operating column temperature, (5) adequate wetting properties compatible with the surface being coated, and (6) a suitable solubility in common volatile solvents.

Various stationary phases can be coated on glass and fused-silica capillaries with different degrees of success. Columns with high numbers of theoretical plates per meter can be prepared with any liquid phase, provided that the optimum treatments, coating solutions, and other necessary parameters discussed in Chapter 2 are properly chosen. The importance of stationary phase selectivity is partially balanced by the large number of theoretical plates obtained in high-resolution work. The high efficiencies of glass WCOT columns usually more than compensate for the selectivities of specific phases, and most separations can be performed on relatively few stationary phases. Thus, in choosing a stationary phase

TABLE 1.9
Common Stationary Phases Used for Coating Capillary Columns

Name	Chemical composition	Polarity $(\Sigma\Delta I)^a$	Operating temp. °C
Oils			
Squalane	C-30 hydrocarbon	0	−40 to 120
OV-101, SP 2100	Methyl silicone oil	229	10 to 260
OV-7	20% phenyl, methyl silicone	592	20 to 300
OV-17	50% phenyl, methyl silicone	884	20 to 300
OV-25	75% phenyl, methyl silicone	1175	20 to 300
OV-210	50% trifluoropropyl, methyl	1520	20 to 210
OV-225	25% cyanoethyl, 25% phenyl, methyl silicone	1813	50 to 220
Silar 5 CP	50% cyanopropyl, 50% phenyl	2428	50 to 240
Silar 10C	100% cyanopropyl silicone	3682	80 to 240
Greases and elastomers			
Kovats phase	C-87 branched hydrocarbon	71	40 to 220
Apiezon L	Hydrocarbon mixture	143	20 to 220
SE-30, OV-1	Methyl silicone gum	216	45 to 320
SE-52	5% phenyl, methyl silicone gum	334	20 to 320
SE-54	1% vinyl, 5% phenyl, methyl silicone gum	337	20 to 320
SP-2125	2% cyanopropyl, 5% phenyl, methyl silicone gum		20 to 300
Ucon LB550	Polyethylene/propylene glycol	496	−20 to 160
Pluronic 64	Polyethylene/propylene glycol		20 to 220
Ucon HB 5100	Polyethylene/propylene glycol	1706	20 to 200
Carbowax 20M	Polyethylene glycol MW 100000	2308	60 to 250
Superox-0.1	Polyethylene glycol gum		60 to 280
Superox-4	Polyethylene glycol gum		60 to 300

^{a} McReynolds constant $(\Sigma\Delta I)$

for relatively simple samples, factors such as performance and thermal stability are more important than selectivity.

For the preparation of highly efficient and inert nonpolar WCOT columns, the formation of a smooth and homogeneous film of stationary phase on an inert and nonadsorbing surface is necessary. Glass is usually wettable by most nonpolar liquids without previous surface modifications. However, adjustment of the glass capillary tubing surface energy to the proper value can result in better wettability and more thermostable surfaces.

Practical experiences in the coating of glass WCOT columns have demonstrated that gum elastomers give columns of consistently higher quality. This is a result of their ready wettability on glass surfaces. Thermostability and resistance to droplet formation are enhanced by the cross-linkages between polymer chains, making the phase viscosity high, even at elevated temperatures. Grob[30] pointed out the benefits of gum phases and concluded that whenever possible a gum phase should be selected. With the recent development in gum elastomers having higher temperature limits and giving a plate number 20–30% higher than liquids, more and more are used for coating of WCOT columns. To keep the column free from moisture and dust, a constant, dry carrier gas flow rate should be maintained. When a WCOT column is not in use, the column should be sealed with a small positive pressure of the dry carrier gas and stored in a dark place.

REFERENCES

1. K. Grob, Plenary Lecture, Eighth Annual Symposium on the Analytical Chemistry of Pollutants, Geneva, April 1978.
2. A. J. P. Martin, *Vapour Phase Chromatography* (D. H. Desty, ed.), Butterworths, London, 1957, p. 2.
3. M. J. E. Golay, *Gas Chromatography 1958* (D. H. Desty, ed.), Butterworths, London, 1958, p. 36.
4. L. R. Snyder, *Principles of Adsorption Chromatography*, Chromatographic Sci. Series, Vol. 3, Marcel Dekker, New York, 1968.
5. G. Dijkstra and J. De Goey, in *Gas Chromatography* (D. H. Desty, ed.), Academic, London, 1958, p. 56.
6. K. Grob, *J. High Res. Chrom. and Chrom. Comm.* 1(5), 263–267 (1978).
7. K. Tešařik and M. Novotny, in *Gas Chromatographie* (H. G. Struppe, ed.), Akademie-Verlag, Berlin, 1958, pp. 575–584.
8. F. I. Onuska and M. E. Comba, *J. Chromatogr.* 126, 133–145 (1976).
9. F. I. Onuska and M. E. Comba, *Chromatographia* 10, 498–503 (1977).
10. G. Schomburg, R. Dielmann, H. Borwitzky, and H. Husmann, *J. Chromatog.* 167, 337–354 (1978).
11. K. Grob, *Chromatographia* 8(9), 423–433 (1975).

12. G. Alexander and G. A. F. M. Rutten, *J. Chromatogr.* **99**, 81–101 (1974).
13. A. I. M. Keulemans, *Gas Chromatography*, 2nd ed., Reinhold, New York, 1959.
14. R. E. Kaiser, *Gas Phase Chromatography*, Vol. 2, Butterworths, London, 1963.
15. M. Novotny and A. Zlatkis, *Chromatogr. Rev.* **14**, 1–44 (1971).
16. M. Novotny, *Anal. Chem.* **50(1)**, 16A–32A (1978).
17. L. S. Ettre, *Open Tubular Columns in Gas Chromatography*, Plenum, New York, 1965.
18. W. Jennings, *Gas Chromatography with Glass Capillary Columns*, Academic, New York, 1978.
19. M. Verzele and P. Sandra, *J. High Res. Chrom. and Chrom. Comm.* **2**, 303–311 (1979).
20. J. H. Purnell, *J. Chem. Soc.*, 1268 (1960).
21. R. E. Kaiser, *Z. Anal. Chem.*, **189**, 1–9 (1962).
22. R. E. Kaiser and R. Rieder, *Chromatographia* **10(8)**, 455–465 (1977).
23. J. C. Giddings, *Dynamics of Chromatography*, Vol. 1. Marcel Dekker, New York, 1965.
24. G. Guiochon, *Anal. Chem.* **50(13)**, 1812–1821 (1978).
25. A. B. Littlewood, *J. Gas Chromatogr.* **1**, 16 (1963).
26. C. E. Figgins and T. H. Risby, *J. Chromatogr. Sci.* **14**, 453–476 (1976).
27. R. D. Schwartz and R. G. Mathews; *J. Chromatogr.* **112**, 111–120 (1976).
28. R. V. Golovnya and T. A. Misharina, *J. High Res. Chrom. and Chrom. Comm.* **3**, 4–10 (1980).
29. R. V. Golovnya and T. A. Misharina, *J. High Res. Chrom. and Chrom. Comm.* **3**, 51–61 (1980).
30. K. Grob, *Chromatographia* **10**, 625 (1977).

GLASS WCOT COLUMNS

The terminology that will be used throughout this chapter will refer to the subject as wall-coated open tubular (WCOT) columns, expressing the fact that the openness of these columns is their most important characteristic.[1] Further, the term WCOT is now used almost universally to specify columns in which the liquid phase is spread directly on the inner wall of a glass capillary tube without the addition of particles that might be considered solid support. Columns in which the coating is deposited on a surface that has been considerably extended by macro-particle deposits during the drawing process are known as porous layer open tubular (PLOT) columns. The term support-coated open tubular (SCOT) columns refers to columns in which the wall has been coated with a mixture of finely divided microparticles, usually silicagel and alumina and a known amount of a liquid phase.[2] Comprehensive reviews have been published.[3,4]

2.1. GLASS SURFACES AS SUPPORT MATERIALS

The commercially available glasses employed for fabrication of WCOT columns contain silicon dioxide moieties (SiO_2) as the major component. This basic building block of all silicate glasses has a silicon atom surrounded by four tetrahedrally located oxygen atoms. However, there is still considerable controversy over the exact structure of even the simplest glass-fused silica.

2.1.1. Fused Silica

A three-dimensional network of SiO_4 tetrahedra exhibiting a random network and lacking any periodicity of symmetry has been postulated as shown in Fig. 2.1.[5]

Fused silica possesses many attractive properties but has one major disadvantage, an inconveniently high melting point. This makes drawing of capillary tubing difficult. The term fused silica denotes material made by fusing synthetic silica produced by burning SiH_4 or $SiCl_4$ in oxygen. This material is significantly lower in metallic impurities than the naturally

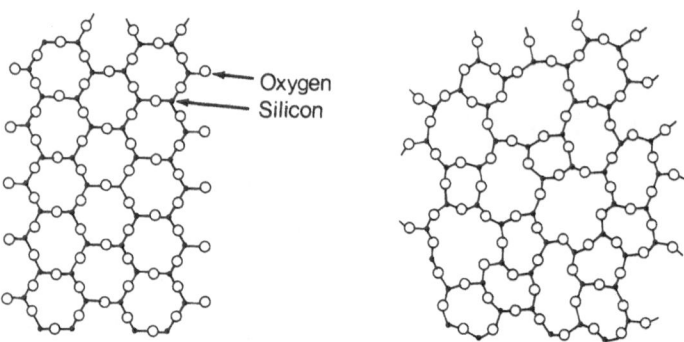

FIGURE 2.1. Two-dimensional representation of crystalline quartz and fused quartz or silica assuming a random network.

occurring quartz, having SiO_2 content greater than 99.9%. Drawn capillary tubing exhibit the high tensile strength of a pristine silica surface. However, a protective coating must be applied to the outer surface to prevent loss of strength caused by the imperfection of the surface and microcracks formed during the drawing processes. The coating must be thermally stable to provide mechanical protection for the column over the entire operating temperature. Its low bleed is also important because the column is inserted into the detector body. Thermal decomposition products from the coating may be swept through the detector and would appear as a baseline offset that resembles a stationary phase bleed signal.

2.1.2. Soft Glasses

The composition of soft glasses is depicted in the schematic two-dimensional arrangement of a typical multicomponent glass shown in Fig. 2.2.

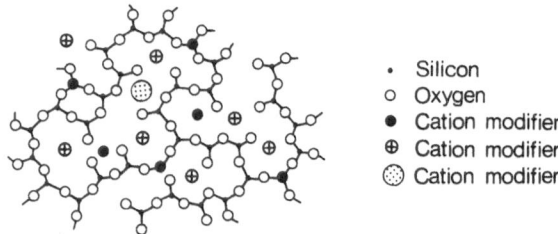

· Silicon
○ Oxygen
● Cation modifier
⊕ Cation modifier
◉ Cation modifier

FIGURE 2.2. Two-dimensional representation of a typical multicomponent glass.

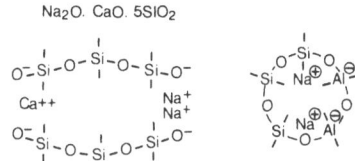

FIGURE 2.3. Essential features of a soda–lime glass.

Typical components include Na$_2$O, K$_2$O, and stabilizers such as CaO, MgO, and Al$_2$O$_3$. Schematic arrangements of soft glass are shown in Fig. 2.3. Sodium oxide contributes to the disruption of Si—O—Si bonds and "softens" the glass.

Sodium oxide contributes further to decrease the viscosity of the glass, increase its solubility in aqueous media and its coefficient of thermal expansion, and lowers its durability. Due to this fact, calcium oxide and magnesium oxide are added to sodium silicate melt to decrease its solubility and to make it more chemically durable.[7] Significant contributions to the construction techniques for capillaries from soda-lime glass has been achieved by researchers.[8-12]

2.1.3. Borosilicate Glass

The most familiar of the borosilicate glasses are known under their trade names of Pyrex, Simax, Kimax, Hysil, and Duran 50. Their composition is broken down as 80–85% silica, 12–14% borax, 4% soda, and 2% alumina.[13-15]

Boron oxide (B$_2$O$_3$) enters the silica network as shown in Fig. 2.4. The boron atom is surrounded by only three oxygen atoms in trigonal coordination. When this is incorporated into the silica tetrahedral arrangement, it has the effect of softening the glass. The addition of alumina as Al$_2$O$_3$ to alkali silicate glasses increases their viscosity, increases their resistance to denitrification, and enhances their chemical durability.

However, there is evidence that the components of a multicomponent glass are not uniformly distributed throughout the glass. It is evident with all these glasses that when they are heated over their softening point they tend to separate in two or more phases having emulsionlike characteristics.

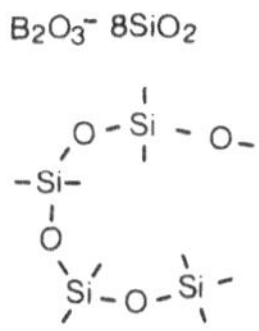

FIGURE 2.4. Essential features of a boro-silicate glass.

TABLE 2.1
Bulk Glass Composition Examples

	Type of glass					
	Pyrex			Potash–soda–lead		
Component	7740	RG flint	Unihost	Uranium 3320	0120	Fused silica
SiO_2	81.0	68.0	69.0	76.0	56.0	99.9
Na_2O	4.0	15.5	18.0	4.0	4.0	
CaO	0.5	6.0	5.5			
ZnO						
Al_2O_3	2.0	3.0	4.0	3.0	2.0	
B_2O_3	13.0			14.0		
BaO		1.0				
MgO		4.0	3.0			
K_2O		0.5	1.5	2.0	9.0	
U_2O_3				1.0		
PbO					290	

In general, glass capillaries drawn from the borosilicate glasses are less fragile than geometrically similar soft glass capillaries. Soda–lime soft glass capillaries have surfaces which are alkaline in nature due to the high content of Na_2O, while the borosilicate glass surfaces are somewhat acidic as a result of the B_2O_3 phase.

The bulk compositions of a number of glass compositions used in the construction of glass capillaries for gas chromatography are presented in Table 2.1.

Poor quality capillaries are drawn from uranium glass.[6] The uranium glass capillaries were very adsorptive, causing tailing of amines, alcohols, and phenols. On the other hand potash–soda–lead glass capillaries usually perform well with phenols and free volatile fatty acids amines, and other active compounds, but sulfur-containing compounds are strongly adsorbed and may even decompose on the lead glass surface.

2.2. GLASS SURFACE PROPERTIES

Considerable progress in the characterization of glass surfaces has been achieved through a number of specific surface analytical techniques such as scanning electron microscopy (SEM), secondary ion mass spectrometry (SIMS), infrared reflection spectrometry, energy dispersive X-ray photoelectron spectroscopy (EDAX), and Auger electron spectroscopy (AES).

These techniques have contributed significantly to the knowledge that the composition of glass surfaces are considerably different from the bulk composition.[16-18] The term bulk refers to those regions of a glass that are sufficiently far from the surface to be unaffected by the specific properties of the surface. It is from these surface layers that important molecular interactions arise affecting the WCOT column preparation and chromatographic performance of these columns.

In general, glass is considered an inert substance in regard to its adsorptive properties and catalytic activity. However, in WCOT column gas chromatography such activity is evident when polar functional groups are present on the compound to be separated. This phenomenon is observed as tailing of a peak, and in severe circumstances by its complete adsorption or decomposition. These interactions between a particular component and the capillary wall are especially noticeable on thin-film WCOT columns, where the degree of shielding provided by the liquid phase is minimal. Surface activity is especially undesirable when picogram quantities of components are being separated.

Capillary wall activity is attributed to the silica surface structure and to impurities present in the surface layers of the glass matrix. The various metallic oxides added during the manufacturing process that are present on or near the surface of the glass can act as Lewis acid sites.[19-21]

These sites are cationic in nature, and the positive charge belongs to a cation of small radius, while the negative charge is distributed over the internal bonds of the incomplete silicate tetrahedra.[22] Lewis acids function as adsorption sites for lone-pair donor molecules such as amines and ketones. Sodium and potassium cations are weaker Lewis acids than magnesium and calcium, which are in turn weaker than boron and aluminum cations. Olefins and aromatic compounds containing π electrons also interact with Lewis acid sites. Boron impurities in silica provide surface Lewis acid sites that are capable of chemisorbing electron-donating molecules.[23] The absence of these adsorption sites on fused-silica surfaces is thought to provide a higher degree of inertness than soda-lime, soft, or borosilicate glasses.

The surface of silica and adsorption on that surface has been investigated.[24-28] Unfortunately, the silica surface of glass has not been studied so thoroughly, and in many cases it is assumed that the silica surface of glass behaves similarily to porous silica.

The single most important structural detail of the silica surface is the distribution of hydroxyl groups that are attached to the surface silicon atoms. These silicon atoms are tetrahedrally coordinated to three other oxygen atoms and hence to the bulk silica as shown in Fig. 2.5.

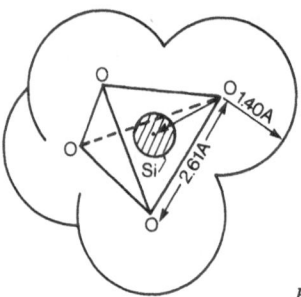

FIGURE 2.5. Basic structure of silica and siliceous glasses.

This structure leads one to infer that at low temperatures, the surface silicon atoms prefer to complete their coordination requirements by attachment to hydroxyl groups rather than by formation of strained siloxane bridges or charged species.[29] The following groups have been found on silica surfaces:

Isolated Silanol   Interacting Silanols   Geminal Silanol   Siloxane

In addition to surface hydroxyl groups there are also hydroxyl groups in the subsurface layers, which are usually termed intraglobular hydroxyls,[30] and some small amount of silanol groups ranging in number between 1 and 2 per 100 Å^2. The relative numbers of the various silanol groups depends on the conditions to which the surface has been exposed. A fully hydroxylated surface contains 1.4 isolated silanol groups, 3.2 interacting silanol groups, and 1.5 geminal silanols per 100 Å^2. Whether two adjacent hydroxyl groups are bound or free is determined by the distance of one hydroxyl group from the oxygen atom of the adjacent hydroxyl group. Hydroxyl groups which are separated from adjacent oxygen atoms by more than 3.1 Å appear to be incapable of hydrogen bonding.[24] By assuming one hydroxyl group per surface silicon atom, there are approximately eight groups per 100 Å^2. Experimental determinations, however, indicate that the surface hydroxyl concentration corresponds to about 5 OH groups per 100 Å^2.[31-33]

2.3. DEHYDRATION AND REHYDRATION OF GLASS SURFACES

Rehydration of fused silica containing 3% B_2O_3 has been studied.[34] The significant observation is that rehydration of the anhydrous surface is followed rapidly by adsorption of water molecules on the newly formed silanol surface. When the fused-silica surface is dehydrated at 800 °C, rehydration is much slower than when dehydroxylated at a lower temperature but can still proceed.

When a completely hydroxylated glass surface is heated, interacting silanol groups begin to condense, thereby eliminating water at about 170 °C. The dehydration is a nonequilibrium process; the rate at which water is removed from the surface at any stage is a function of temperature and the concentration of remaining silanol groups. The dehydration is reversible up to about 400 °C but is irreversible and is completed at about 800 °C. This process is depicted in schemes 1 and 2:

$$\equiv Si - O \diagup^{H} \quad \xrightleftharpoons{400\ °C} \quad \begin{matrix} \equiv Si \\ \equiv Si \end{matrix} \diagdown O + H_2O \qquad (2.1)$$

$$\equiv Si - O \diagup^{H} \quad \xrightarrow{800\ °C} \quad \begin{matrix} \equiv Si \\ \equiv Si \end{matrix} \diagdown O + H_2O \qquad (2.2)$$

Water molecules are adsorbed only on the hydroxylated silica surface and not on the siloxane surface, which is essentially hydrophobic.[35] It may be illustrated by the following data[36] showing that the total surface energy of silica of different known surface areas have different enthalpies:

Pure siloxane surface	259 ± 3 erg cm^{-2}
Pure silanal surface	129 ± 8 erg cm^{-2}

and thus the heat of hydration is equivalent to 130 ± 7 erg cm^{-2}. The total energy of the silanol-covered surface is then only slightly higher than the total surface energy of water (118.5 erg cm^{-2}) at ambient temperature.

Wettability of Glass Surfaces. Coating of the solid surface by a liquid is of basic importance in WCOT column technology. To achieve high separation efficiency, a uniform and homogeneous film of stationary phase must coat the inner wall of the glass capillary. This well-defined film must maintain its integrity and not rearrange to form droplets as the temperature is varied.

The nonwetting or hydrophobic character shown by the siloxane surface when contacted with polar liquid phases indicates that the work of adhesion of polar liquid to a glass surface consists of dispersion van der Waals forces and hydration of nonionic polar sites, i.e., bonding with SiOH groups.[37] On silica surfaces hydration controls wettability.

The generally accepted theory for spreading of liquid on solid materials is based on the concept of contact angle.[38] The contact angle θ of a liquid drop on a solid surface is described by:

$$\gamma_{SG} = \gamma_{SL} + \gamma_{LG} \cos \theta \tag{2.3}$$

where γ is the surface tension denoted for all interfaces such as gas, liquid, and solid. The work of adhesion and the surface tensions γ_{SL} exist as the phase boundary of a drop of liquid at rest on a solid surface. When a droplet is placed on a solid surface, it may spread to cover the surface or it may remain as a droplet. The angle between the tangent to the liquid droplet and the solid surface is defined as the contact angle (see Fig. 2.6). When $\theta = 0$, the liquid spreads freely over the surface. As θ increases, the tendency for a liquid to spread decreases.

The wettability of a surface is a thermodynamic function of the equilibrium between the cohesion forces inside the liquid and the energy of the solid surface. The cohesion forces inside the liquid are characterized by the surface tension, and the energetics of the solid surface are characterized by the surface free energy. The contact angle depends upon the specific surface free energy of the liquid and the solid phase. Spreading generally occurs when the specific surface free energy of the liquid is less than that of the specific glass surface. Compositions of both the spreading liquid phase and the solid surface are of primary importance since the surface atoms of both phases are attracted to each other by van der Waals dispersion forces.

The critical surface tension (CST, γ_c) represents the value of the liquid surface tension above which liquids show a finite contact angle on a given surface. The value of the CST is obtained from a plot of $\cos \theta$ versus γ of a homologous series of liquids as shown in Fig. 2.7. This

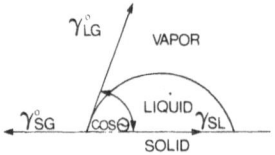

FIGURE 2.6. Contact angle diagram.

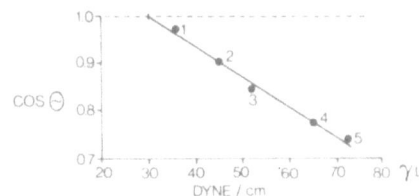

FIGURE 2.7. Contact angle for various
liquid phases on untreated Pyrex.

plot is usually a straight line; the intercept with the cos θ = 1 line gives
the CST value.

Most organic liquid phases exhibit large contact angles on glass and
do not form uniform layers but break up into droplets during the wetting
process.[9] It has been reported[9,39,40] that typical stationary phases are in
the range of 30–50 dyn/cm, and the CST for smooth, clean glasses is 30
dyn/cm. This value explains the poor wettability of glass surfaces. Various
treatments, however, can be employed to raise the CST of the glass.
Surface tensions of some stationary phases and of solvents are given in
Tables 2.2 and 2.3.

In some instances the surface tension of the stationary phase can be
lowered by the addition of a surfactant. Polydimethylsiloxane is adsorbed
on the surface of silica whether the surface is dehydrated or hydroxylated.[41] Polydimethylsiloxane oil of molecular weight 350,000 provide a
film thickness of 10 nm, and one molecule covers 6.10^4 nm^2. This is as
flat as the strained siloxane chain can lie. Its surface tension is in the
range of 19–20 dyn/cm, which is always less than the critical surface
tension of its own adsorbed layers.

TABLE 2.2
Surface Tension (γ) of Stationary Phases at Ambient Temperature

Stationary phase	γ (dyn/cm)	cos θ on Pyrex glass
Squalane	29.95	0
OV-101 (dimethyl silicone)	20.40	
QF-1 (trifluoropropyl silicone)	24.60	
OV-210 (trifluoropropyl silicone)	23.60	
OV-17 (methylphenyl silicone)	30.40	
Apiezon L	33.20	
Ucon 50 HB-2000	35.70	
Polypropylene sebacate	40.20	
OS-124 (polyphenylether)	46.10	
Diethyleneglycol succinate	50.90	
Carbowax 400	44.20	24
Diethylhexyl sebacate	31.10	

TABLE 2.3
Surface Tension (γ) of Solvents

Solvent	γ (dyn/cm)	Solvent	γ (dyn/cm)
n-Pentane	16.0	Acetone	23.3
n-Hexane	18.4	Methanol	22.6
Benzene	28.9	Ethanol	22.3
Dichloromethane	28.1	Diethyl ether	
Tetrachloromethane	25.6	Di-n-propyl ether	20.0

However, many ester-type stationary phases hydrolyze slightly on glass surfaces, and the resulting monolayer has a critical surface tension that is less than the surface tension of the liquid phase itself. Such action is consistent with observation of WCOT columns losing their efficiencies due to film breakup after a few days or even hours of operation.

2.4. CAPILLARY TUBING FABRICATION

The difficulties with the physical adsorption and catalytic activity of metal capillary tubing led Desty[42] to design a special apparatus for drawing and coiling glass capillaries. This concept was adopted by Huppe and Bush Co. and later by Hewlett-Packard and Shimadzu in Japan. With the development of dependable glass drawing technology, the Desty drawing machine has been modified to speed up the drawing process.[43] The preparation of glass WCOT columns can be increased up to 10-fold in drawing speed.[44] It has been reported that nonuniformity of the column inner diameter can effect the performance characteristics of WCOT columns, but a precision of 1.5% can be attained by modifying a commercial glass drawing machine.[45] The basic instrument is shown in Fig. 2.8. In

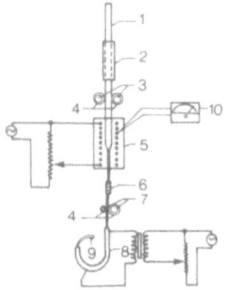

1 Material glass tubing
2 Supporter
3 Feeding rollers
4 Silicone rubber tires
5 Furnace
6 Guide pipe
7 Drawing rollers
8 Coiling tube
9 Supporting rod
10 Pyrometer

FIGURE 2.8. Capillary drawing apparatus (Shimadzu GDM-1).

principle, a glass tubing is driven by means of the feeding rollers into the furnace. The glass tubing softened by the heat is drawn by the drawing rollers; the drawn tubing is then coiled by allowing it to pass through the stainless steel bending tube, which is heated by directly applying electrical power. Changing the ratio of the revolution of the feeding rollers and the drawing rollers, or by using a glass tubing of a suitable inner and outer diameter, a desired inner diameter between 0.1 and 1 mm can be produced.

Fused-silica or quartz capillaries are drawn using advanced fiber optics technology. This material provides a degree of chemical inertness unmatched by other glasses. Moreover, when drawn with a very thin wall and externally coated with a suitable polymer to protect its outer surface, the fused-silica glass tubing exhibits a high degree of flexibility that greatly facilitates its handling as a column in gas chromatographs.

Figure 2.9 shows apparatus employed for the drawing of flexible-walled fused-silica glass capillary tubing having 0.2–0.25-mm i.d. × 0.28–0.33-mm o.d. A vertically mounted high-temperatured furnace which contains high-purity graphite heating elements operates in an oxygen-free atmosphere by employing argon as a scavenger gas.[45]

With this apparatus it is claimed that the permanently coiled thick-wall fused-silica WCOT column tubing does not require an outer coating to protect it from atmospheric corrosion or associated problems in handling. However, this material is not commercially available, and probably it would be easier to coat it with many more liquid phases. In contrast, the thin-walled tubing requires an outer polymeric coating, approximately 50 μm thick, for protection and additional strength. The outer polymeric coating should withstand temperatures in excess of 400 °C which represents a problem to all manufacturers. However, polyimides used as coatings may be used up to 350 °C. The thin-walled coated tubing is inherently straight and flexible material which makes the columns re-

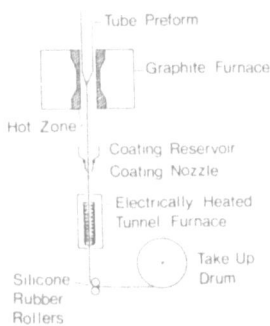

FIGURE 2.9. Apparatus used for drawing fused-silica and natural quartz capillary tubing and the on-line coating of the thin-walled tube with a polymeric sheath.

markably durable. The tubing can be readily inserted by means of graph-itized Vespel or graphite ferrules into the connections of the injector and detector ports of gas chromatographs without difficulty. This character-istic both greatly simplifies the handling of the WCOT column and im-proves chromatographic performance by eliminating unnecessary extra connectors.

Both the glass and fused-silica tubing as originally obtained from the manufacturers invariably contain oil and many impurities. The fused silica also contains scratches and pitting of the inner wall and outer surfaces.

Soft glass tubings contain Ca^{+2} and Mg^{+2} ions which have an adverse effect on the separation of some polar compounds.

In comparing the effect of several leaching treatments to prepare the surface of the raw tubings, it has been found that nitric acid, hydro-fluoric acid washing, and chromic acid–sulfuric acid have a beneficial effect on both the inactivity and the wetting properties of the glass surface.[47] A pretreatment step has been proven essential for the success of WCOT column preparation.[48] As shown in Table 2.4, the critical surface tension of Pyrex glass tubing changes drastically after a pretreat-ment step is applied. This observation indicates that the source of the nonwettability and surface activity does originate from factors in the bulk glass—impurities of external origin which can be removed by some clean-ing treatment.

A procedure has been employed that uses 10% hydrofluoric acid and allows it to stand in a tube for up to 48 hr.[47] Afterward, the tube is filled with concentrated nitric acid and allowed to stand at room tem-perature for 1 hr, after which time the tube is rinsed with HF and distilled deionized water. The resultant surface after draining gave a sintered appearance with evenly distributed pit holes as shown in Fig. 2.10. This surface appears to be better deactivated than the chromic acid–sulfuric acid treated inner wall.

After their drawing, fused-silica capillaries exhibit residual impurities. These impurities are concentrated on the inner wall of the capillary column due to a diffusion process and temperature gradient toward the cooler inner walls. A bulk concentration of 2 ppm sodium can result in

TABLE 2.4
Critical Surface Tension (γc) of Pyrex Glass

Treatment	γc (dyn/cm)
Acetone washed	28
Chromic-sulfuric acid cleaned	44
NaOH etched	33

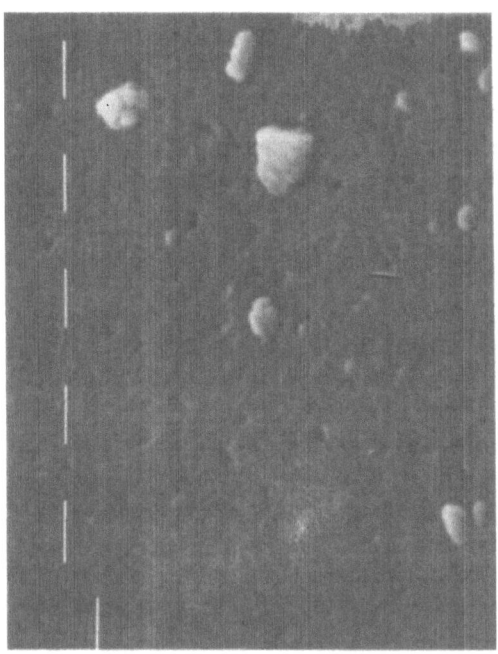

FIGURE 2.10. Sintered surface of
HF–HNO₃ glass capillary.

2% sodium on the surface after annealing at 1700 °C. Since vitreous silica containing 2 ppm sodium is considered extremely pure, it can be seen that in most precipitates, sodium may play an unrecognized role.

A stream of dry argon to flush the hot zone of the tubing during the drawing operation can be used.[49] The gas leaves a clean and smooth surface, and the capillary is wettable by OV-101, UCON-LB, OV-17, and Carbowax 20M. Soft glass and Pyrex glass capillaries in which purified oxygen is used during drawing have been studied by surface analysis with Auger electron spectroscopy and X-ray photoelectron spectroscopy.[50] They showed little or no carbon present in the glass. These data confirmed that leaching Pyrex glass with 20% HCl solution at 110 °C for 48 hr produced a nearly pure silica surface which retains its integrity after heat treatment at 400 °C. Leaching of soft glass was not complete.

2.5. WCOT COLUMN SURFACE MODIFICATION

2.5.1. General Considerations

There are only few liquid phases that will produce a satisfactory thin uniform film on the untreated glass surface since soft Kimble glass or Pyrex glass, including vitreous silica, is nonwettable by many of the common stationary phases. Only methylsilicone oils or gums such as SP-2100, OV-101, OV-1, or SE-30 can be coated successfully on untreated glass, but even these are subject to localized thickening. The more polar stationary phases having a higher surface tension will condense into droplets. The cohesive forces are greater than adhesive or wetting forces between the liquid phase and the glass surface. This causes the liquid phase to retreat from the surface and contract into droplets, thus pre-empting the formation of a thin, uniform layer which is necessary for high-efficiency columns. These problems have been attacked either by changing the wetting properties by addition of surfactants or by changing the characteristics of the glass surface.

A procedure for fabrication of WCOT columns uses a quarternary phosphonium halide such as benzyltriphenyl-phosphonium chloride, which aids in coating the wall of a capillary with substrate and causes it to adhere firmly.[51] Trioctadecyl ammonium bromide may also be used as an additive.[52] These methods have not found too many applications.

2.5.2. Surface-Roughening Techniques

Surface roughening will decrease the apparent contact angle between the glass wall and the liquid phase and thus facilitate its spreading. Such roughening greatly enhances the wettability of the glass surface by the liquid stationary phase. Generally, a liquid will spread better on a rough rather than on a smooth surface because the surface covered by the liquid droplet releases more energy due to interfacial forces. For a rough surface there is more surface area under the droplet and, therefore, more energy is released. The influence of surface roughening becomes apparent by a decrease in the contact angle. According to the Wenzel equation[53]:

$$r = \frac{\cos \theta_1}{\cos \theta_2} = \frac{A_1}{A_2} \tag{2.4}$$

where A_1 and A_2 refer to the microscopic and the macroscopic surface areas, respectively, and θ_1 and θ_2 refer to the apparent contact angles measured on the roughened and the smooth surfaces, respectively. The value of r is close to unity for freshly drawn glasses, but becomes greater than unity as the surface is roughened. Correspondingly, the apparent contact angle becomes smaller.

2.5.3. HCl-Induced Crystallization in Gaseous Phase

The alkali ions in soft glass are rather mobile at elevated temperatures and can react with acids. If the acid is in gaseous phase, the resulting salt will deposit onto the surface as a uniform crystalline layer. This type of surface roughening has been described by a number of research-ers.[54–62] A comprehensive study of the mechanism of the hydrogen chloride soda-glass etching process can be summarized as follows[62]:

1. During the drawing of capillary tubing from a soda-glass, vapor-phase sodium oxide condenses on the inner wall of the tubing during cooling, forming an alkali-rich surface layer representing sites of nucleation.

2. The sodium ions move rather freely through the lattice toward the surface. At the same time hydrogen ions from the hydrogen chloride diffuse into the glass surface.

3. At the surface sodium ions react with hydrogen ions as follows:

$$-Si-ONa + H^+ \longrightarrow -Si-OH + Na^+ \qquad (2.5)$$

4. Sodium ions diffuse through the glass, associate with chloride anions, and randomly locate on a nucleation site.

5. As the hydrogen chloride decreases, particle growth slows down and recrystallization becomes competitive.

6. If the heat treatment is continuing for long periods, large sodium chloride crystals grow at the expense of smaller ones. Marshall and Parker[58] have found that uniform growth is optimized by heating at 360 °C for 3 hr.

In conclusion, the limitations of this method of surface roughening are summarized as follows:

1. The procedure is essentially limited to soft glass columns.

2. Crystal growth depends to a large extent on the surface composition of the glass and, therefore, the reproducibility of an ideal microroughness is difficult.

3. The solubility of sodium chloride in various solvents can present a problem during coating of capillary tubings with the stationary liquid.

4. Capillaries coated with thin films exhibit some adsorption properties resulting from the weak Lewis acid interaction of the sodium ions.

5. High concentration of alkali on the inner surface increases the catalytic decomposition of various liquid phases with resultant bleeding at higher temperatures.

2.5.4. Pyrex Glass Surface Modification

A simple method for the surface treatment of Sial glass, which has composition similar to Pyrex glass employes both HF gas and 2-chloro-

1,1,2-trifluorethyl methyl ether.[63] Although the capillaries contained an opaque deposit, it was not until 1975 that the nature of the deposit was eventually identified under scanning electron microscope as polycrystalline silica in the form of filamentary crystals or, as referred to in literature, "silica whiskers."[64-66] Hydrogen fluoride attacks the silica network forming volatile silicone fluorides. The effect of hydrogen fluoride etching is more complex than that of HCl etching. When the concentration of HF is higher, the formation of silicone tetrafluoride is enhanced. When a system is under nitrogen at high temperature, SiF_4 decomposes at nucleation centra formed on the "kinks" (irregularities on the surface) and "whiskers" start to grow. The filamentary crystal-covered surface has an increased effective area, and it is an advantageous feature of the process.[67] When ammonium hydrogen difluoride is used, the maximum whisker growth occurred after only 3 hr in contrast to 2-chloro-1,1,2-trifluoroethyl methyl ether method requiring 24 hr at 450 °C. Figure 2.11 shows a series of SEM micrograms of a cross section of a column and various types of filamentary crystals which were prepared by etching with 2% ammonium hydrogen difluoride at 450 °C. Using a potassium hydrogen difluoride solution, Heckman[68] produced capillaries from soft and Pyrex glass which turned opaque in a very short time at room temperature. According to the authors, columns were washed with water and dried and were ready for deactivation and coating. In our hands filamentary

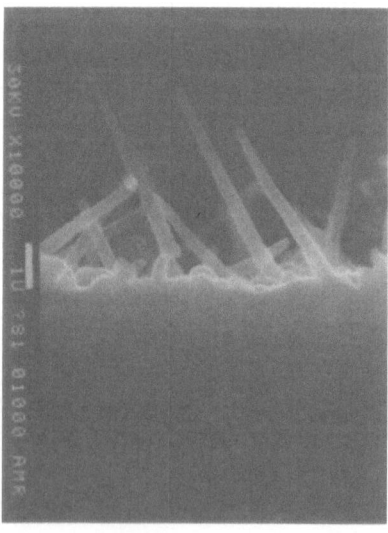

FIGURE 2.11. SEM-gram of various types of filamentary crystals.

crystal WCOT columns are excellent capillaries, and they can be easily deactivated by means of Carbowax 20M or silylation.

2.5.5 Deposition of Barium Carbonate

Perhaps the most used procedure for preparation of WCOT columns consists of dynamically coating the glass surface with barium hydroxide solution employing carbon dioxide gas to push the plug of barium hydroxide solution through the capillary tubing.[69] During this step, barium carbonate crystals grow from nuclei on the glass surface. The structure of a barium carbonate crystal produced on the glass surface is influenced by a large number of experimental parameters including glass surface structure, crystallization temperature, and addition of surfactants.[70,71] Barium carbonate crystals are formed on any kind of glass surface, but the glass surface influences the size, shape, and distribution of the crystals.

On untreated soft glass smaller and less distant particles are formed than on untreated Pyrex, borsilicate glass. These differences become smaller with prior acid leaching.

Three specific procedures have been outlined for the formation of barium carbonate deposits. By varying the concentration of the barium hydroxide solution, different degrees of surface coverage are possible. Saturated barium hydroxide produces thick layers of barium carbonate that are suitable for coating with polar stationary phases. For less polar phases a 1:10 dilution of the barium hydroxide is used. Finally, for the preparation of apolar columns, a 1:100 dilution of the barium hydroxide is used. Schematic presentation of the process is shown in Fig. 2.12.

Although the deposition of barium carbonate increases the glass surface area and therefore improves the wettability of the surface, especially for polar stationary phases, there are some problems concerning surface activity. It was found that the glass surface after barium carbonate deposition is slightly basic, and that acids with pK_a values lower than 6

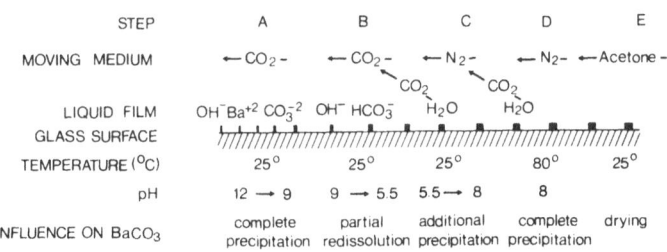

FIGURE 2.12. Schematic presentation of the deposition of $BaCO_3$ process.

have difficulty passing over the carbonate surface. Furthermore, it was found essentially impossible to deactivate satisfactorily barium carbonate crystals which were not covered by liquid phase unless a Carbowax deactivation after barium carbonate deposition is used.[69] It has become apparent that the barium carbonate procedure is principally a surface-roughening technique, and it is somewhat improper to refer to the technique as one which also provides surface deactivation.

2.5.6. Other Roughening Methods

There are several procedures used to deposit carbon black inside the capillary column by pyrolyzing hydrocarbons.[72] The deposition of a carbon layer by means of the pyrolysis of methylene chloride is one of the more effective methods, but reproducibility and uniformity of the carbonized surface caused problems. Stable carbon layers have been prepared on the inner glass capillary walls by dynamic[73–75] and static coating procedure[76] using a colloidal solution of graphitized carbon black. The resultant layer strongly adheres to the glass surface and is not removed by washing it with different polarity solvents. The graphitized carbon-black-coated columns permit the use of a wide range of stationary phases, while those prepared by the hydrocarbon pyrolysis method are suitable only for a restricted range of moderately polar phases.

A dynamic method for depositing suspended fine particles of silanized silic acid uses Silanox 101 in the stationary phase as it is coated on the inner wall of the capillary column. The two-step dynamic method is useful in the preparation of nonpolar columns only.

Since the hydrophobic surface of Silanox is not wettable with polar stationary phases, it would seem logical to use an unsilanized fused silica known as Cab-O-Sil in much the same way that Silanox has been used.[77] Cab-O-Sil is employed as a stabilizer for the polar OV-225 stationary phase.

A low-temperature plasma-etching method, performed at low pressure, uses fluorine-containing compounds under the RF (radiofrequency) discharge to form a plasma that reacts with the glass wall.[78] It is thought that the principal etching mechanism represents the reaction of fluorine radicals with the silica of the glass surface.

A procedure for preparing glass WCOT columns for liquid chromatography where a layer of silica gel is formed by the reaction between a solution of ammonia and tetramethylammonium hydroxide can be used.[79] The surface layer prepared with 4% tetramethylammonium hydroxide has a silica gel layer resembling a wrinkled cloth (see Fig. 2.13).

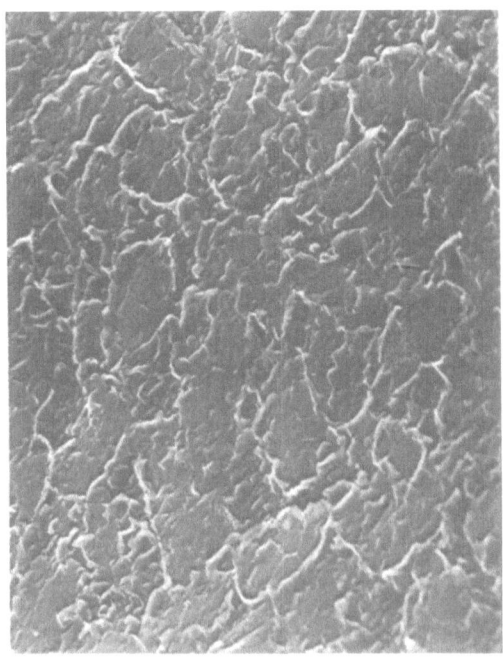

FIGURE 2.13. Silica layer resembling a wrinkled cloth.

2.6. GLASS SURFACE DEACTIVATION AND CHEMICAL MODIFICATION

The surface modifications discussed earlier were designed mainly to provide better wettability of the glass surface with respect to the stationary phase. Attention has been directed also toward deactivation of the surface.

The glass surface has several different types of active sites originating from various ingredients or from the silica structure itself. When the treatment of increasing the wettability is achieved, it is usually followed by a deactivation step. In principle, the surface energy has to be first increased to achieve complete wetting, and thereafter it should be decreased again to eliminate active centers which exhibit a very high energy on the glass surface. To clarify the nature of active centers is a difficult task since the exact surface composition and structure of glass surfaces is not fully understood.[80]

Earlier, interest was focused on the surface silanol groups ($\equiv$Si— OH),[81] and most research activities were focused on elimination of these groups from the surface. Later, it became evident that the surface silanol groups were not the only active sites, but that di- and trivalent ions, such as calcium, magnesium, aluminum, boron, and iron, also act as active centers. The elimination of these elements from the surface can be achieved by acid leaching, which can be performed with relative ease. Surface silanol groups cannot be removed by this method. They may be almost completely eliminated by blocking the surface silanols by means of chemical modification.

During acidic leaching the basic and alkali oxide components are dissolved to yield a silica-rich surface layer. Below pH 7, the original glass backbone of siloxane (Si—O—Si) bonds is unaffected. Alkaline solutions break the Si—O—Si bonds and by nucleophilic attack from silanol (Si—OH) and Si—O$^-$ bonds, thus bringing about the actual dissolution of the silica surface. Rather than an enriched surface layer of silica, a porous surface structure is formed.

It has been observed that an acid leaching would be useful for the removal of metallic cations from the soft glass.[82] At the same time acid leaching removes the cations and gives improved chromatographic performance.[83] It has been reported that soft glass capillaries washed with formic acid were superior to those which had not been acid treated.[84]

A similar procedure has been reported but with NaCl crystals removed from the glass surface by washing with water and rinsing with acetone and diethyl ether followed by HF diluted with nitrogen.[85] This treatment forms a thin layer of dealkalinized glass similar to fused silica. It leads to increased temperature stability of the coated stationary phase.

Dynamic leaching of both soft and Pyrex glasses has been used.[86,50] In this method 20% HCl is slowly forced through the glass capillary tubing at 100 °C. Auger electron spectra of the interior surface leached in this way show no distinguishable amount of boron and other metallic impurities. In fact, depth profile analyses indicate that a relatively thick layer of silica surface is formed. Thus, a leached glass surface of this nature has many of the same properties that give fused silica its inertness. Under static conditions it is possible that boron ion reprecipitation could take place. The other ions extracted from the surface could also be reprecipitated. As the pH of the static leaching solution increases from the alkali extracted into it, the reaction kinetics could change from diffusion controlled to interface controlled. When this happens the silica-rich surface region starts to depolymenize and is eroded away.[87] The original barium carbonate procedure has been modified to aleviate some of the problems associated wih the leaching step and to compensate for excessive surface dehydroxylation at 180 °C and shorter lengths of time.[88]

Although the surface concentration of silanol groups, and therefore activity of a capillary tubing, can be decreased by thermal treatment,[89] they can be more effectively eliminated by removing active hydrogens through a chemical modification, some being more effective than others. The most widely applied methods of chemical surface modification in glass WCOT column preparation is silanization.

In this reaction the surface hydroxyl groups of the glass are replaced with silyl ether groups. The polarity and chemical characteristics of the modified glass surface that is formed by silylation can be controlled by the choice of the silylation reagent. Silanization of glass capillaries was first reported using trimethylchlorosilane.[90,91] The reaction can be formulated as

$$Si_sOH + X - Si(CH_3)_3 \longrightarrow Si_s - O - Si(CH_3)_3 + HX \qquad (2.6)$$

It is known that at lower temperatures methylchlorosilanes are physically adsorbed on the silica surface and are not chemically reacted.[92] It would appear that reaction temperatures of 300–400 °C are necessary to ensure complete reaction.

Deactivation of glass capillary surfaces by silanization by means of different regents such as dimethyldichlorosilane, trimethylchlorosilane, hexamethyldisilazane, and N, O-bis-(trimethylsilyl) acetamide in different solvents under various reaction conditions have been studied.[93] It was found that the effectiveness of the silanization procedure can be improved if the treatment is performed with hexamethyldisilazane at 300 °C for a period of about 20 hr.

With hexamethyldisilazane (HMDS) the overall reaction taking place with the glass surface is.

$$(CH_3)_3Si - NH - Si(CH_3)_3 + 2Si_s - OH \longrightarrow 2Si_s - O - Si(CH_3)_3 + NH_3$$

$$(2.7)$$

It has been determined that the reaction of HMDS with a silica surface follows second-order kinetics.[94] In other words, when one hexamethyldisilazane molecule reacts with the surface, two hydroxyl groups are removed. Data indicates that about 70% of the HMDS molecules are reacting on two sites, and 30% react by chemisorption on a single site.[95] HMDS is considerably more reactive toward the surface silanols than chorosilanes with chemisorption taking place at a lower temperature.

The bonded silyl groups are extremely stable and may be heated up to 500 °C in vacuum or 400 °C in air without decomposition.[96] It has been demonstrated that silanization with a gaseous mixture of HMDS and trimethylchlorosilane (TMS) improved coatings wih nonpolar stationary phases.[97-99] A variety of differently substituted silanes were used to form surfaces wettable with medium polarity siloxanes.

A procedure in which the capillary column is heat treated at 400 °C to provide better surface silanization and the use of diphenyltetramethyldisilazane to provide a more stable surface layer than HMDS has been suggested.[100]

It has been shown that silanization at 400 °C is more complete than that obtained at lower temperatures.[50] Dynamic gas phase silanization is used to ensure excess regent and to minimize reaction product interferences. The degree of surface coverage by silanization is related to the number of free hydroxyl groups on the surface. Layer amounts of silane could be bonded if a glass wall was previously corroded by acid treatment.[101]

Significant progress in a deactivation of glass and fused-silica surfaces is accomplished by creating a nonextractable layer of polymethylsiloxane phase (SP-2100, SE-52, and SE-54) on the glass surface.[102] Employing the dynamically coating technique, the capillary is filled with nitrogen and sealed on both ends, and then heated to 450 °C for 2–20 hr. During heat treatment, chemical bonding takes place between the methylsiloxane moiety and the surface silanol groups. After heat treatment, the column is rinsed with solvent to remove nonbonded residue. The column is then recoated with the same stationary phase to produce a film of defined thickness and polarity. Reports of the preparation of capillary WCOT columns containing chemically bonded silicone phases were reported as early as 1968.[103-105] Chemical bonding has been shown to increase the stability of the stationary phase layer as compared with the conventionally coated layers. The synthesis of siloxane polymers from a mixture of two differently substituted silanes provides a method for the manufacture of highly stable glass WCOT columns of accurately controlled polarity.[106]

In situ syntheses of methyl and phenyl polysiloxanes is done by dynamically coating the capillary with silicone tetrachloride, followed by a solution of the polysiloxane polymer formed by hydrolysis of mixtures of dimethyldichlorosilane and methyltrichlorosilane.[107,108] The column was then sealed and heated at 320 °C for 20 hr. This treatment produces a moderate amount of cross-linking in addition to chemical bonding to the glass surface, which has been suggested as a means to increase the film stability.[109]

As a result of the previous treatments, the glass capillary tubing surface energy should be adjusted to the proper value suitable for spreading a liquid stationary phase on the surface (Table 2.5). The object of this treatment is the deposition of a thin, uniform, well-defined film of stationary phase throughout the capillary column, usually 0.05–0.5 μm

TABLE 2.5
Surface Tensions (γ) of Common Stationary Phase Solutions

Stationary phase	γ (dyn/cm)	Temp (°C)	Solvent	(w/w)	(v/v)	(w/w)
PEG 400	23.4	25	Acetone	10		
PEG 400	25.5	25	Acetone	20		
DC-560	22.9	25	Acetone	12		
DC-560	23.3	25	Acetone	20		
SF-96	24.5	25	Toluene		2.0	
SF-96	25.0	25	Toluene		10.0	
SF-96	24.9	22	Toluene		10.0	
Apiezon L	25.1	22	Cyclohexane		12.8	10.0
SE-30	25.6	22	Chloroform		5.9	5.4
SE-30	25.7	22	Chloroform		3.6	3.3
OV-17	28.7	22	Toluene		10.0	
Carbowax 20M	26.9	22	Chloroform		5.0	5.0

thick. This is necessary in order to obtain the highest possible separation efficiency and resolution-including capacity of the column. Most of the stationary phases are viscous liquids or gums. They are dissolved in a suitable solvent. It is advantageous to use freshly prepared, well-degassed solutions since it was found that the siloxane-based phases undergo decomposition in solution in a relatively short time.[110]

2.7. COATING THE WCOT COLUMN

2.7.1. Dynamic Methods

The dynamic coating procedure consists in filling 5–15 coils of the capillary column with a solution of the stationary phase, followed by forcing this solution through the capillary column at a linear velocity of 1–2 cm/sec with dry nitrogen or helium.[111] A thin film of this solution is deposited on the capillary wall. Continual flushing with the carrier gas after coating evaporates the remaining solvent and leaves a thin layer of stationary phase.

The advantage of the method compared to a static coating procedure discussed below is that it is fast. A disadvantage is that the amount of stationary phase on the capillary wall is not precisely known. The amount of solution left behind a moving plug depends on a number of factors.[112] They are the surface tension, the viscosity of the coating solution, and the diameter of the capillary column. Another problem arises from the fact that the solution film before the evaporation of the solvent is too thick and its viscosity is too low to resist the smooth evaporation under gas flow conditions. As a result, secondary plugs are formed, which results in a nonuniform coating of the capillary wall. It is possible to summarize the reasons for nonuniform film formation as follows:

1. As the coating solution is removed from the capillary, the coating velocity increases, which results in a thicker film at that end of the capillary column. A buffer column (approximately one-half of the length of the coated column) is often attached to the end of the capillary to avoid this problem.
2. Since the amount of coating solution decreases with the length of the capillary column, it results in an increase of the linear velocity of the plug as coating proceeds and, therefore, an increasing film thickness is formed.
3. The solvent evaporation involves transport of some of the stationary phase toward the end of the capillary column, which results in increasing film thickness along the column.[112,113]
4. Temperature fluctuations along the capillary column cause the solvent to distil from the warmer part of the column and condense at the cooler part. This causes film nonuniformity and droplet formation.[114]
5. Poor wettability of the glass surface by either the stationary phase solution or the stationary phase itself may result in droplet formation and subsequent nonuniform film of a stationary phase. The instability of a coating of liquid due to axial spreading out on the inside of a capillary column results in formation of lenses which is independent of surface tension and viscosity and equals $2\pi r/0.7$, where r is the radius of the tube.

The wettability problem is very dependent on the critical surface tension of the glass surface, and the surface tensions of the stationary phase coating solution and the stationary phase itself.

Solutions of stationary phases generally have lower surface tensions than the pure stationary phases by higher surface tensions than the pure solvents. The actual surface tensions of a stationary solution can be calculated by the additive equation:

$$\gamma = \gamma_1 x + \gamma_2(1 - x) = \gamma_2 + (\gamma_1 - \gamma_2)x \tag{2.8}$$

where γ, γ_1, and γ_2 are the surface tensions of the mixture and of components 1 and 2, respectively, and x is the proportion of component 1 expressed in (w/w).

As stated in Sec. 2.3 for uniform spreading of the stationary phase, the critical surface tension of the surface must be greater than the surface tension of the liquid. During the coating of a capillary column, the stationary phase and the solvent compete for adsorption on the active sites. The relative proportions of surface covered by each are dependent not only on the relative proportion of solute and solvent but also on their relative adsorptivities.[115]

Treatment with n-pentane, methylene chloride, or acetone reduces the critical surface tension of glass up to 50%.[117] Furthermore, it is suspected that the coating solvent is not completely removed during column conditioning, and that the effects of these solvents are still present.

The formation of droplets can be largely avoided, or at least greatly reduced, by controlling the coating speed,[114] concentration of the coating solution,[116] solvent volatility,[117] temperature,[114] and rate of solvent evaporation.[113]

Methylene chloride has been a very popular solvent for the dynamic coating procedure because of its low boiling point and nonflammability. However, the high vapor pressure of methylene chloride increases lenses and droplet formation during the waiting process. The problem may be eliminated by temperature programming the capillary column in a water bath from 20 to 32 °C during coating and up to 42 °C during the solvent evaporation. Maintaining a carrier gas flow through the capillary column for several hours after coating reduces the formation of droplets.

The most significant development in the dynamic coating procedure was the introduction of the mercury plug method.[118,119] This method involves adding a drop of mercury between the solvent plug and the carrier gas, which wipes most of the coating solution off the surface as the plug moves through the capillary column. More concentrated solutions are used in this procedure, resulting in the formation of films that resist drainage during the drying process. The mercury plug method eliminates the disadvantages mentioned above, and it is likely to be one of the standard procedures in WCOT glass capillary column chromatography. A general equation for predictions of the film thickness in dynamically coated capillary columns is[120]:

$$d_f = \frac{1.34rc}{100} \frac{(\mu\eta)^{2/3}}{\gamma} \tag{2.9}$$

This equation is very close to the equation proposed by Novotny and colleagues[97] because, when a narrow range of flow velocity is investigated, there is not much difference between a dependence on the power of 0.5 or the power of 0.67.

The (d_f) film thickness in dynamically coated capillary columns depends directly on the radius of the capillary, where u is the average velocity of the coating plug, c is the concentration of the solution used (in v/v %), and η is the viscosity of the solution and γ its surface tension. The general formula proposed is:

$$d_f = \frac{rc}{200} \frac{(\mu\eta)^{1/2}}{\gamma} \qquad (2.10)$$

This equation can be used to predict the average film thickness to be obtained when preparing capillary columns from any solution when η and γ have been measured and u is controlled. It should be taken into account that both η and γ depend on the concentration of the solution used.

The values of η and γ required for the calculation are generally not available. They can, however, be obtained by simple measurements using an Ubbelohde viscometer and the capillary rise method. The type of equation that best describes the film thickness in the dynamic method is applicable also to the mercury plug method.[121]

A simple method can be used for calculating the film thickness. A few coils of the column are coated with a plug of solution of a known length, and the shortening of the plug over a given column length is measured. This procedure is repeated several times with increasing coating speed each time. The film thickness at these different coating speeds can be easily calculated from the concentration of the solution, the amount of solution used over a given length, and the inner diameter of the capillary column. A graph can be constructed of the coating speed versus film thickness, from which the coating speed can be selected to give a desired film thickness.

It is recommended that after the coating solution and the mercury plug are discharged from the capillary column, the flow of carrier gas is continued while the temperature of the capillary column is increased to remove residual solvent. Afterward, the column outlet is connected to a detector.

2.7.2. Static Techniques

In the static method the column is completely filled with a dilute solution containing from 1 to 15 mg/ml of a stationary phase in a suitable low-boiling solvent, and one end is carefully sealed. It is absolutely critical that the capillary be completely filled, with no trace of air at the sealed end. After closing one end, the solvent is evaporated through the open end by connecting it to a vacuum pump. The solvent evaporates under quiescent conditions, and a thin film of stationary phase remains on the capillary wall. This leaves a thin film of stationary phase, the thickness (d_f) of which can be calculated from:

$$d_f = \frac{r}{2\beta} = \frac{dc}{400} \quad [\mu m] \qquad (2.11)$$

where r is the radius of the capillary and β is the phase ratio or where d is the capillary diameter in μm and c is the concentration of a liquid phase in the volume percent.[122] To calculate the phase ratio from a weighed amount of stationary phase, the density of the stationary phase must be known (Table 2.6).

A simple setup for the static coating of a glass capillary tubing is shown in Fig. 2.14. This procedure consists in connecting one end of a short piece of Teflon-shrinkable tubing or silicone rubber tubing to the end of the capillary column which is to be closed for solvent evaporation under vacuum at room temperature. The other end of the Teflon is drilled through the body of the clamping device, as is evident from the Fig. 2.13. By turning the two units of the clamping device together, the Teflon tubing is squeezed and vacuum-tight closure of the capillary is achieved. The other end of the glass capillary to be coated is connected to an evacuated (20-torr) 4-l round glass two-necked flask into which the solvent is evaporated. The connection is made by means of a silicone rubber tubing. Fluctuations in room temperature are eliminated by keep-

TABLE 2.6
Specific Gravity of Common Stationary Phases

Stationary phase	s.w. (g/cm³)	Stationary phase	s.w. (g/cm³)
OV-1	0.980	SE-30	0.960
OV-3	0.997	SE-54	0.980
OV-7	1.021	SF-96-200	0.972
OV-11	1.057	SF-96-2000	0.974
OV-17	1.092	DC-200	0.970
OV-22	1.127	DC-510	1.000
OV-25	1.150	DC-550	1.068
OV-61	1.090	DC-710	1.100
OV-101	0.975	PEG 400	1.125
OV-105	0.990	DEGS	1.260
OV-210	1.320	QF-1	1.320
OV-225	1.086	Silar 5CP	1.125
OV-275	1.160	Silar 10C	1.116
OS-124	1.210	SP 2401	1.300
Squalane	0.830	XE 60	1.080
AN600	1.080	Apiezon	

FIGURE 2.14. Apparatus for the static procedure according to Husman and Schomburg.

ing the capillary column under a suitable beaker. The most popular solvent for the static coating has been methylene chloride because of its volatility and excellent solvating properties.

The use of n-pentane for the coating of apolar gum phases is superior to other solvents including methylene chloride because evaporation can be accomplished in approximately half the time necessary for methylene chloride solutions.[123] The time necessary to coat a capillary column having a given length and 0.2-mm i.d. is shown in Fig. 2.15.[124]

Another method involves screwing the open end of the filled capillary column into a high-temperature oven to force solvent evaporation.[125] When all but the last few coils have been driven into the oven, the sealed end is broken open, and the residual solvent vapors are removed, either by suction or by dry nitrogen. The WCOT columns prepared by this technique are not evenly coated, and thus they are not of superior quality. Examination shows that at regular intervals, thicker annular deposits of liquid phase occur. The problem may be diagnosed as due to the solvent not evaporating evenly. While one end of the capillary column is open and the capillary column has a certain mass, it relatively slowly reaches the boiling point temperature, but very quickly reaches the boiling point when heat transfer inside the heating coil abruptly superheats the solvent. It evaporates in bursts, and distances between bursts is variable. A mod-

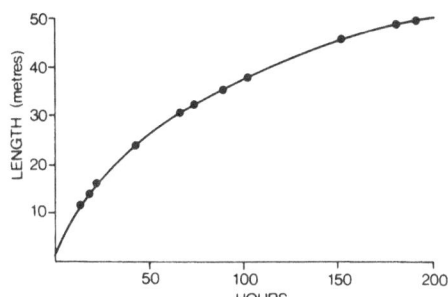

FIGURE 2.15. Variation of the time necessary for static coating of glass capillary column having 0.2-mm i.d. and containing methylene chloride at 22 °C.

ified version of the above method uses an arc of stainless steel tubing, heated electrically to a high temperature at the oven entrance.[126] As the capillary column enters the oven through the hot entrance tube, the solvent evaporates, instantaneously forming a superheated vapor which also contains an aerosol of the stationary phase. A significant pressure is created, up to 30 atm, as this solvent evaporation continues because of the spontaneous introduction of the solution-filled capillary column into the entrance tube. The inner wall of the capillary column is well cleaned by high-temperature and high-pressure aerosol spraying, which also contains the stationary phase. The column wall may be partially dehydrated by this treatment, resulting in a better contact between the capillary column wall and the stationary phase. This may help to explain the increased film stability on the inner wall surface.

Organic peroxides producing free radicals have been very widely used as initiators and cross linking agents. Of the peroxides most commonly used benzoyl peroxide decomposes thermally to give two benzoyloxy radicals. Both of the radicals are capable of initiating polymerization, and it has been shown that this indeed happens. However, peroxides generally show some tendency to undergo side reactions with other radicals in the system. Consequently, the order of decomposition is not quite unity.

We suggest using azo compounds instead of peroxides. These compounds, possessing the general formula,

$$R-N=N-R$$

decompose unimolecularly with the elimination of nitrogen. They are not subject to any side reactions as far as is known and apparently decompose at the same rate independent of the solvent used. For cross-linking, we prefer 1-*tert*-butylazo-1-cyanocyclohexane which decomposes at 96 °C or azobis-isobutyronitrile which decomposes at 103 °C.

A cross-link may be looked upon as a more or less permanent connection between two polymer molecules, binding them together through a system involving primary chemical bonds. A new cross-linked polymer will no longer be completely soluble under the usual conditions because the macromolecules are bonded to each other and cannot be completely separated by solvent. Instead, the solvent will simply cause the macromolecule to swell to some limiting value which is proportional to the amount of cross-linking.

A coating of fused-silica WCOT columns by the modified static coating procedure according to Onuska[46] is described in detail below.

In this technique the fused-silica capillary tubing is filled with a coating solution containing an appropriate concentration of a liquid phase, preferably silicone gums, Apiezon grease, or Superoxes, and 0.4–1.0% 1-*tert*-butylazo-1-cyanocyclohexane (LUAZO-96, Penwalt Corp.) in *n*-pentane. A short length (less than 5 cm) of one end of the capillary is not filled with the solution. This end is sealed and at least 40 cm of the other end of capillary is also left empty. This open end is later clamped to the metallic cylinder which is placed on a horizontal shaft driven by an electric dc motor geared to vary the shaft speed from 1 to 5 revolutions per minute (Fig. 2.16). The shaft can be demounted easily from the motor by a bayonet-type connector. The drum is immersed into a silicone oil

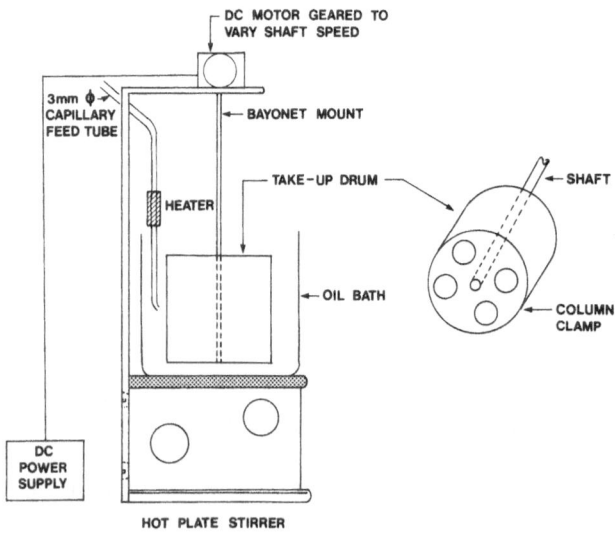

FIGURE 2.16. Apparatus for the static coating procedure of fused-silica columns according to Onuska.

bath, which is located on a stirrer hot plate. Proper adjustment permits the capillary tubing to be driven smoothly through the silica tube heater, which consists of a concentric high-temperature resistor (250 Ω; 5 W) positioned over a silica tube that is well isolated by asbestos tape. The bath temperature is raised to 120 °C, the tube heater to 300 °C, and then the drive motor is energized. The evaporation runs continuously. The process can be easily monitored visually to see fine bubbles of the evaporating solvent leaving the capillary. Instantaneous evaporation of the evaporating solvent is very even since the oil bath provides a pressure head which eliminates vigorous bumping. Approximately 60 min are required to coat a 50-m WCOT column.

The static coating apparatus consists of a hollow aluminum cylinder 855 mm long × 114.5 mm in diameter, having wall thickness 3 mm with a stainless-steel shaft down the center fixed at both ends of the cylinder. One end of the shaft, which protrudes 45 mm beyond the cylinder, has a thumb screw mount. Attached to the shaft, inside the cylinder, near each end is an expanding spider (Fig. 2.17).

One arm of each spider is connected to one of the three grooved Lexan strips, allowing their diameter to be adjusted from that of the aluminum cylinder to 12.5 mm larger to properly fit and hold the glass capillary column. To mount a WCOT column made from glass or fused

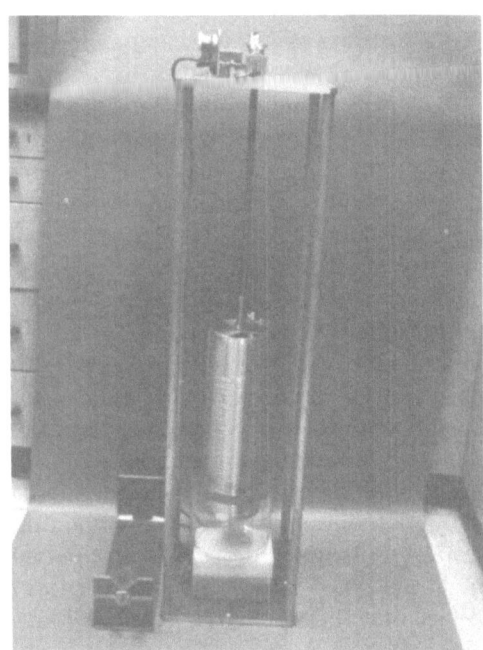

FIGURE 2.17. Universal apparatus for the static coating according to Onuska.

silica, the Lexan strips are retracted and the column is slipped over the cylinder, which is then set horizontally in a holding stand. The spiders are expanded until the Lexan strips protrude above the surface of the cylinder sufficiently so that the column loops can be positioned into each groove, or every other groove of the Lexan strips depending on the spacing desired. Each Lexan strip has one hundred and sixty 2.5-mm deep $\times$ 1-mm wide grooves spread 2 mm apart, center to center.

Once the capillary is positioned, the cylinder is mounted to the rack gear, which is attached to a 990-mm tall stand. A constant-speed timer motor attached to a pinion driving gear moves the rack gear downward at 2 cm/min, immersing the column at a uniform rate into a heated silicone oil bath. The motor has an automatic shut-off switch and is connected to the rack and pinion gear through a bayonet clutch, which allows the rack gear and cylinder to move manually when required.

The most important advantages of this technique over classical static coating procedures are the speed of coating and the ability to apply any stationary phase. The coating apparatuses are relatively inexpensive and may serve both the coating and cross-linking purposes at the same time.

2.8. WCOT COLUMN CONDITIONING

A freshly coated stationary phase retains traces of solvent, oligomers that are always present in the polymeric material of the stationary phase and in those cases where heat has been applied to the capillary column, e.g., during straightening the ends after coating, it will also contain degradation products of the stationary phase. These components pass through the WCOT column and emerge at the detector to yield an unstable baseline and artifacts. A conditioning step is necessary to remove such substances.

It is necessary to condition older columns from time to time. This conditioning helps to remove the accumulation of slow-moving components from previous injections in the column. It can also result if the operator has neglected the molecular sieve filters in the chromatograph when concentrating carrier gas impurities in the WCOT column. A degree of compromise is necessary in selecting the conditioning temperature. High conditioning temperatures (above 220 °C) achieve stable baselines more rapidly, but they shorten the life of any WCOT column. Selection of the conditioning temperature should be considered in a view of the proposed analyses, detector to be employed, the temperature limits of the liquid phase and purity, and dryness of the carrier gas. It is a matter of fact that the upper temperature limit of a stationary phase on glass surface is lower than that for packed columns. If the proposed analyses involve temperatures of 80–150 °C, there is no reason to condition the column at 200 °C. Overnight conditioning at normal flow rates at 175 °C may be sufficient and will not be affecting the WCOT column life. Especially when an electron capture detector is used, it is desirable to disconnect the WCOT column from EC detector or N–P detector during conditioning because the detector will not be contaminated with impurities and lower molecular weight oligomers. If the WCOT column is left connected, the detector should be maintained at a maximum recommended temperature, which should be at least 50 °C over the WCOT column temperature limits. Even more frequent detector cleaning will be required at times.

New commercially obtained columns and in-house made columns should be subjected to normal carrier gas flow rates at 100 °C for 30–60 min. Afterward, they may be programmed (e.g., 4 °C/min) to the conditioning temperature and held. All WCOT columns are quickly deteriorated by exposure to high temperatures with no flow or very restricted flow of carrier gas.

2.9. GLASS WCOT COLUMN MAINTENANCE

The ideal glass WCOT column is generally considered to possess a high separation efficiency and excellent deactivation with an appropriate temperature stability. As is shown in the text, the stationary phase distribution as a thin uniform film over the inner surface is responsible for it. WCOT columns that have been subjected to a large number of splitless injections; the stationary phase displacement is usually observed in one or two loops at the inlet end of the WCOT column. Removing the defective part of the WCOT column will usually restore efficiency.[127]

It is evident that a method for manufacturing glass WCOT columns plays an important role in determining the lifetime of WCOT columns. Methylsilicone oil such as SP-2100 columns deactivated with Carbowax 20M have an upper temperature limit of 240 °C, while those deactivated by silanization or high-temperature silanization with SP-2100 have an upper limit of 320 °C, especially on the fused-silica surfaces. Capillary columns coated with Carbowax 20M will withstand up to 1 year at column temperature not exceeding 170 °C. At temperatures higher than 220 °C their lifetime drops to 1 month, and at 250 °C they rarely last more than 1 week.

The useful WCOT column life is defined as the time in which the column loses half of its amount of stationary phase.[128] Columns coated with a dimethyl polysiloxane phase CP-Sil-5 can tolerate limited exposures to temperatures as high as 350 °C.[129] The methyl phenyl silicones, such as SE-52 and SE-54, on a well-deactivated glass surface or fused-silica columns are capable of tolerating long-term exposures at 300 °C. Cyanopropyl silicone liquid phases can be used as high as 260 °C, but it depends upon the polarity of a liquid phase as can be seen in Table 2.7.

It should be emphasized that temperature resistance varies from manufacturer to manufacturer, and that without exception, lower temperatures are conducive to longer capillary column life. The column life is also influenced by the purity and dryness of a carrier gas. Air is usually injected with the sample and ppm concentration of oxygen is usually present in helium or nitrogen. Similarily, moisture is usually present in the carrier gas. These components must be removed from carrier gas by the installation of oxygen traps and driers in the carrier gas line and by careful handling of the injected samples. Acetone, also acetonitrile, car-

TABLE 2.7
Stationary Phases Commonly Used for WCOT Columns

Name	Chemical nature	Polarity	Temperature range (°C)
Squalane	C-30 hydrocarbon	0	−40–120
Kovats hydrocarbon	C-87 branched alkane	71	40–220
Apiezon L	Alkanes	143	25–220
SE-30, OV-1	Dimethyl silicone gum	216	25–350
SE-52	5% phenyl, methylsilicone	334	25–320
SE-54	1% vinyl, 5% phenyl, methylsilicone	337	25–300
OV-7	20% phenyl, methylsilicone	592	25–300
OV-17	50% phenyl, methysilicone	884	25–280
OV-25	75% phenyl, methylsilicone	1175	25–280
OV-210, QF-1	50% trifluoropropyl, methylsilicone	1520	25–200
OV-225	25% cyanoethyl, 25% phenyl, methylsilicone oil	1813	50–210
OV-275	100% cyanopropyl silicone oil	4938	100–250
Ucon LB550	Polyethylene-propylene glycol	496	−20–180
Pluronic 61	Polyethylene-propylene glycol		0–200
Emulphor 0	Polyethylene glycol-octadecyl ether	1587	20–220
Ucon HB5100	Polyethylene-propylene glycol	1706	20–240
Pluronic 64	Polyethylene-propylene glycol		20–240
Carbowax 20M	Polyethylene glycol	2308	65—240
Superox-0.1	Polyethylene glycol, M.W. 100,000		60—280
Superox-4	Polyethylene glycol, M.W. 4M		60–300
Dexil 300	Carborane, methylsilicone	474	20–350
Dexil 400	Carborane, methyl phenylsilicone	587	20–350
Poly-S 179	Polyphenylether sulfone		200–400

bon disulfide, low molecular weight alcohols, and tetrahydrofuran must be avoided as a solvent.

Research in this area is progressing quite rapidly. The lifetime of glass WCOT columns and the maximum operating temperatures are being improved both by the production of more stable stationary phases and decreased catalytic activity of the glass surface.

REFERENCES

1. L. S. Ettre, *Open Tubular Columns in Gas Chromatography*, Plenum, New York, 1965.
2. I. Halasz and E. Heine, *Anal. Chem.* **37**, 495–500 (1965).
3. V. Pretorius and J. C. Davidtz, *J. HRC & CC* **2**, 703–711 (1979).
4. M. L. Lee and W. W. Wright, *J. Chromatogr.* **184**, 235–312 (1980).
5. W. H. Zachariasen, *J. Am. Chem. Soc.* **54**, 3841 (1932).
6. R. Dandeneau and E. Zerenner, *J. HRC & CC* **2**, 351–356 (1979).
7. W. M. Novotny and A. Zlatkis, *Chromatogr. Rev.* **14**, 1–44 (1971).
8. K. Tesařik and M. Novotny, in *Gas Chromatographic 1968* (H. S. Struppe, ed.), Akademie-Verlag, Berlin, 1969.
9. G. Alexander and G. A. F. M. Rutten, *J. Chromatogr.* **99**, 81–101 (1974).
10. G. Schomburg and F. Weeke, *Gas Chromatography 1972* (S. G. Perry, ed.), Institute of Petroleum, London, 1973, p. 285.
11. M. L. Lee, K. D. Bartle and M. Novotny, *Anal. Chem.* **47**, 540–543 (1975).
12. K. Grob, *Helv. Chim. Acta* **51**, 718–737 (1968).
13. K. Grob and S. Grob, *J. Chromatog. Sci.* **8**, 635–639 (1970).
14. F. I. Onuska, B. K. Afghan and R. J. Wilkinson, *J. Chromatogr.* **158**, 83 (1978).
15. V. Pretorius, J. W. du Toit, and J. C. Davidtz, *J. HRC & CC* **4**, 79–80 (1981).
16. S. M. Budd, in *Glass Surfaces* (D. E. Day ed.), North-Holland, Amsterdam, 1975, p. 55.
17. D. M. Sanders and L. L. Hench, *J. Am. Ceram. Soc.* **52**, 666 (1973).
18. D. M. Hercules, *Anal. Chem.* **50**(8), 734A–744A (1978).
19. N. W. Cant and L. H. Little, *Can. J. Chem.* **42**, 802–809 (1964).
20. A. M. Filbert and M. L. Hair, *J. Gas Chromatogr.* **6**, 218 (1968).
21. I. D. Chapman and M. L. Hair, *Trans. Faraday Soc.* **61**, 1507 (1965).
22. A. V. Kiselev, *J. Phys. Chem.* **38**, 1501–1508 (1964).
23. N. W. Cant and L. H. Little, *Can. J. Chem.* **43**, 1252–1254 (1965).
24. V. Ya. Davidov, A. V. Kiselev, and L. T. Zhuravlev, *Trans. Faraday Soc.* **60**, 2254 (1964).
25. F. H. Hambleton, J. A. Hockey, and J. A. E. Taylor, *Trans. Faraday Soc.* **62**, 801 (1966).
26. M. L. Hair, *Infrared Spectroscopy in Surface Chemistry*, E. Arnold, London, 1967.
27. J. B. Peri and A. L. Hensley, *J. Phys. Chem.* **72**, 2926–2933 (1968).
28. M. L. Hair, *Glass Surfaces* (in D. E. Day, ed.), North-Holland, Amsterdam, 1975, p. 301.
29. L. T. Zhuravlev, A. V. Kiselev, V. P. Naidina, and A. L. Polyakov, *Russ. J. Phys. Chem.* **37**, 1216 (1963).
30. V. Y. Davydov, A. V. Kiselev, V. A. Lokutsievskii, and V. I. Lygin, *Russ. J. Phys. Chem.* **47**, 460 (1973).
31. L. T. Zhuravlev and A. V. Kiselev, *Russ. J. Phys. Chem.* **39**, 236 (1965).

32. R. K. Iler, *The Chemistry of Silica*, Wiley-Interscience, New York, 1979.
33. K. Unger, *Porous Silica*, Elsevier, Amsterdam, 1979.
34. V. R. Dietz and N. H. Turner, *J. Phys. Chem.* **75**, 2718–2727 (1971).
35. G. J. Young and T. P. Bursh, *J. Colloid Sci.* **15**, 361 (1960).
36. S. Brunauer, D. L. Kantro, and C. H. Weise, *Can. J. Chem.* **34**, 1483 (1956).
37. J. Laskowski and J. A. Kitchener, *J. Colloid Surfce Sci.* **29**, 670 (1969).
38. W. A. Zisman, in *Adhesion and Cohesion* (P. Wels ed.), Elsevier, New York, 1962.
39. K. Tesařik and M. Nečasova, *J. Chromatogr.* **65**, 39–46 (1972).
40. M. Nečasova and K. Tesařik, *J. Chromatogr.* **79**, 15–22 (1975).
41. A. V. Kiselev, V. N. Novikova, and Y. A. Eltekov, *Dokl. Akad, Nauk SSSR* **149**, 131–137 (1963).
42. D. H. Desty, J. N. Haresnape, and B. H. F. Whyman, *Anal. Chem.* **32**, 302–304 (1960).
43. E. A. Mistryukov, L. S. Savichev, and V. O. Gavrilov, in Proc. 2nd Danube Symposium, Carlsbad, Czechoslovakia, 1979, *Czechosl. Chem. Soc.*, **B.2.21** (1979).
44. J. G. Scheuing, L. G. J. van der Ven, and A. Venema, *J. HRC & CC* **1**, 101–102 (1978).
45. S. R. Lipsky, W. J. McMurray, M. Hernandez, J. E. Purcell, and K. A. Billeb, *J. Chromatogr. Sci.* **18**, 1–9 (1980).
46. F. I. Onuska, Proceedings of the 18th International Symposium Advances in Chromatography 1982, Tokyo, Japan, April, 1982, pp. 15–16.
47. W. G. Jennings, K. Yabrumoto, and R. H. Wohleb, *J. Chromatogr. Sci.* **12**, 344 (1974).
48. J. Simon and L. Szepesy, *J. Chromatogr.* **119**, 495–504 (1976).
49. B. W. Wright, M. E. Lee, S. W. Graham, L. V. Phillips, and D. M. Hercules, *J. Chromatogr.* **119**, 355–369 (1980).
50. E. J. Malec, *J. Chromatogr. Sci.* **9**, 318-320 (1971).
51. L. D. Metcalf and R. J. Martin, *Anal. Chem.* **39**, 1204–1205 (1967).
52. R. N. Wenzel, *Ind. Eng. Chem.* **28**, 988 (1936).
53. K. Tesařik and M. Novotny, in *Gas Chromatographie 1968* (H. G. Struppe, ed.), Akademie-Verlag, Berlin, 1968, p. 575–584.
54. G. Alexander and G. A. F. M. Rutten, *Chromatographia* **6**, 231–233 (1973).
55. G. Alexander and G. A. F. M. Rutten, *J. Chromatogr.* **99**, 81–101 (1974).
56. G. Alexander, G. Garzo, and G. Palyi, *J. Chromatogr.* **91**, 25–37 (1974).
57. J. L. Marshall and D. A. Parker, *J. Chromatogr.* **122**, 425–442 (1976).
58. H. T. Badings, J. J. G. van der Pol, and J. G. Wassink, *Chromatographia* **8**, 440–448 (1975).
59. J. Krupčik, M. Kristin, M. Valachovičova, and S. Janiga, *J. Chromatogr.* **126**, 147–160 (1976).
60. D. H. Parker and J. L. Marshall, in Proc. 2nd Intern. Symposium on Glass Capillary Chromatography, (R. E. Kaiser, ed.), Hindelang, Germany, p. 299–332, 1977.
61. J. J. Franken, G. A. F. M. Rutten, and J. A. Rijks, *J. Chromatogr.* **126**, 117–132 (1976).
62. M. Novotny and K. Tesařik, *Chromatograhia* **1**, 332–333 (1968).
63. J. D. Schieke, N. R. Comms, and V. Pretorius, *J. Chromatogr.* **112**, 97–107 (1975).
64. F. I. Onuska and M. E. Comba, *J. Chromatogr.* **126**, 133–145 (1976).
65. P. Sandra and M. Verzele, *Chromatograhia* **10**, 419–425 (1977).
66. F. I. Onuska and M. E. Comba, in Proc. 2nd Intern. Symposium on Glass Capillary Chromatography (R. E. Kaiser, ed.), Hindelang, Germany, pp. 283–298, 1977.
67. R. A. Heckman, Ch. R. Green, and F. W. Best, *Anal. Chem.* **50**, 2157–2158 (1978).
68. K. Grob and G. Grob, *J. Chromatogr.* **125**, 471–485 (1976).
69. K. Grob, G. Grob, and K. Grob, Jr., *Chromatographia* **10**, 181–187 (1977).
70. K. Grob, J. R. Guenter, and A. Portman, *J. Chromatogr.* **147**, 111–117 (1978).
71. K. Grob, *Helv. Chim. Acta* **51**, 718–737 (1968).

72. G. Nota, G. C. Goretti, M. Armenante, and G. Marino, *J. Chromatogr.* **95**, 229–231 (1974).
73. G. C. Goretti, A. Liberti, and G. Nota, *Chromatographia* **8**, 486–490 (1975).
74. G. C. Goretti, A. Liberti, and G. Pili, *J. HRC & CC* **1**, 143–148 (1978).
75. C. Vidal-Madjar, S. Bekassy, M. F. Gonnord, P. Arpino, and G. Guiochon, *Anal. Chem.* **49**, 768–772 (1977).
76. A. L. German and E. C. Horning, *J. Chromatogr. Sci.* **11**, 76–82 (1973).
77. C. A. Cramers, E. A. Vermeer, and J. J. Franken, *Chromatographia* **10**, 412–418 (1977).
78. A. Y. Masada, K. Hashimoto, T. Inone, Y. Sumida, T. Kishi, and Y. Suwa, *J. HRC & CC* **2**, 400–404 (1979).
79. K. Tesařik, *J. Chromatogr.* **191**, 25–30 (1980).
80. M. Novotny and A. Zlatkis, *Chromatogr. Rev.* **14**, 1–44 (1971).
81. M. Novotny and K. D. Bartle, *Chromatograhia* **7**, 122–127 (1974).
82. J. C. Diez, M. V. Dabrio, and J. L. Oteo, *J. Chromatogr. Sci.* **12**, 641–646 (1974).
83. M. L. Lee, K. D. Bartle, and M. Novotny, *Anal. Chem.* **47**, 540–543 (1975).
84. H. Borwitzky and G. Schomburg, *J. Chromatogr.* **170**, 99–124 (1979).
85. M. L. Lee, D. L. Vassilaros, L. V. Philips, D. M. Hercules, H. Azumaya, J. W. Jorgenson, M. P. Maskarinec, and M. Novotny, *Anal. Lett.* **12**, 191–203 (1979).
86. L. L. Hench, *J. Non-Cryst. Solids* **25**, 343 (1977).
87. K. Grob, G. Grob, and K. Grob, Jr., *Chromatographia* **10**, 181–187 (1977).
88. K. Grob, G. Grob, and K. Grob, Jr., *J. HRC & CC* **2**, 677–679 (1979).
89. A. V. Kiselev, *Gas Chromatograhy 1962* (M. van Swaay, ed.), Butterworths, London, 1962.
90. A. V. Kiselev and K. D. Shcherbakova, *Gas Chromatographie 1962* (M. Schroeter and K. Metzner, eds.), Akademie-Verlag, Berlin, 1962, pp. 207 and 241.
91. F. O. Stark, O. K. Johannson, G. E. Vogel, R. G. Chaffee, and R. M. Lacefield, *J. Phys. Chem.* **72**, 2750–2754 (1968).
92. Th. Welsch, W. Engewald, and Ch. Klaucke, *Chromatographia* **10**, 22–24 (1977).
93. M. L. Hair and W. Hertl, *J. Phys. Chem.* **73**, 2372–2378 (1968).
94. W. Hertl and M. L. Hair, *J. Phys. Chem* **75**, 2181–2185 (1971).
95. R. Evans and T. E. White *J. Catal.* **11**, 336 (1968).
96. M. Novotny, L. Blomberg, and K. D. Bartle, *J. Chromatogr. Sci.* **8**, 390–393 (1970).
97. M. Novotny and K. D. Bartle, *Chromatographia* **3**, 272 (1970).
98. M. Novotny and K. D. Bartle, *Chromatographia* **7**, 122–127 (1974).
99. K. Grob, G. Grob, and K. Grob, Jr., *J. HRC & CC* **2**, 31–35 (1979).
100. M. H. van Rijswick and K. Tesarik, *Chromatographia* **7**, 135–137 (1974).
101. G. Schomburg, H. Husmann, and H. Borwitzky, *Chromatographia* **12**, 651–660 (1979).
102. C. J. Bossart, U.S. Pat. 3,514925, 1970.
103. J. Jonsson, J. Eyem and J. Sjoquist, *Anal. Biochem.* **51**, 204–219 (1973).
104. C. Madani, E. M. Chambaz, M. Rigaud, J. Durand, and P. Chebroux, *J. Chromatogr.* **126**, 161–169 (1976).
105. C. Madani and E. M. Chambaz, *Chromatographia* **11**, 725–730 (1978).
106. L. Blomberg, J. Buijten, J. Gawdzik, and T. Wännman, *Chromatographia* **11**, 521–525 (1978).
107. L. Blomberg and T. Wännman, *J. Chromatogr.* **168**, 81–88 (1979).
108. K. Grob, *Chromatographia* **10**, 625 (1977).
109. A. Venema, L. G. J. van der Ven, and H. van den Seege, *J. HRC & CC* **2**, 69–70 (1979).
110. G. Dijkstra and J. De Goey, in *Gas Chromatography 1958* (D. H. Desty, ed.), Academic, New York, 1958, p. 56.
111. K. D. Bartle, *Anal. Chem.* **45**, 1831–1836 (1973).

112. L. Blomberg, *Chromatographia* **8**, 324–326 (1975).
113. J. Roeraade, *Chromatographia* **8**, 511–516 (1975).
114. H. W. Fox, E. F. Hare, and W. A. Zisman, *J. Colloid Sci.* **8**, 194 (1953).
115. L. Blomberg, *J. Chromatogr.* **138**, 7–16 (1977).
116. D. A. Parker and J. L. Marshall, *Chromatographia* **11**, 526–533 (1978).
117. G. Schomburg, H. Husmann and F. Weeke, *J. Chromatogr.* **99**, 63–79 (1974).
118. G. Schomburg and H. Husmann, *Chromatographia* **8**, 517–530 (1975).
119. G. Guiochon, *J. Chromatogr. Sci.* **9**, 512 (1971).
120. G. Alexander and S. R. Lipsky, *Chromatographia* **10**, 487–491 (1977).
121. G. A. F. M. Rutten and J. A. Rijks, *J. HRC & CC* **1**, 279 (1978).
122. K. Grob, *J. HRC & CC* **1**, 93 (1978).
123. J. Merle d'Aubigne, C. Landault, and G. Guiochon, *Chromatographia* **4**, 309–312 (1971).
124. E. L. Ilkova and E. A. Mistryukov, *J. Chromatogr. Sci.* **9**, 569–570 (1971).
125. W. G. Jennings, K. Yabamoto, and R. H. Wohleb, *J. Chromatogr. Sci.* **12**, 344–348 (1974).
126. F. Berthou and Y. Dreano, *J. HRC & CC* **2**(5), 251–252 (1979).
127. K. Grob and H. J. Jaeggi, *Chromatographia* **5**, 382–391 (1972).
128. E. Thizon, C. Eon, P. Valentin, and G. Guiochon, *Anal. Chem.* **48**, 1861–1865 (1976).
129. W. J. M. Houtermans and C. P. Boodt, *J. HRC & CC* **2**(5), 249–250 (1979).

INLET SYSTEMS

3.1. GENERAL CONSIDERATIONS

Ideally a sample mixture is introduced quantitatively into a carrier gas stream. It should be introduced without decomposition and move into the capillary column as a homogeneous unbroadened plug so that its volume is a minimum under the inlet conditions. However, vaporization of sample, especially if it is contained in a volatile solvent, disturbs the gas flow and produces a temperature drop from the absorption of the heat of vaporization.

High-resolution gas chromatography injection systems contain critical pneumatic components, especially the pressure-regulated system with either manual or automatic flow controllers. All of the active pneumatics components are thermostated to better than ± 1 °C for maximum retention time reproducibility and detector baseline stability. Stable, resettable, low flow rate controllers have been developed for capillary gas chromatographs that are superior to using fixed restrictors as a quasimass flow controller.[1] A WCOT column system must be designed with pneumatics for carrier gas and auxilliary gas lines with accurate metal bellow and metal diaphragm pressure controllers, pressure gauges, and possibly metal on-off valves for an easier and faster flow rate measurement and leak checks. The carrier gas circuit must also contain in addition to the pressure controller, high-precision flow controllers for constant-flow operations during temperature-programming operation. It is mandatory that this system be well thermostated.

The schematic flow diagram for a WCOT column gas chromatographic operation is shown in Fig. 3.1. The pneumatic system consists of a pressure regulator and a makeup valve for supplying additional gas to a detector (such as an electron capture detector), a pressure gauge indicating the inlet pressure to the injector–splitter, and a series of traps and filters to remove contaminants from the gas lines. It is imperative that the pressure regulators and makeup valves are installed in the temperature-controlled compartment.

The injector–splitter is designed to be used as a conventional splitter, a splitless injector or a direct injector, depending on the configuration of the glass insert employed. The injector–splitter (Fig. 3.2) is enclosed

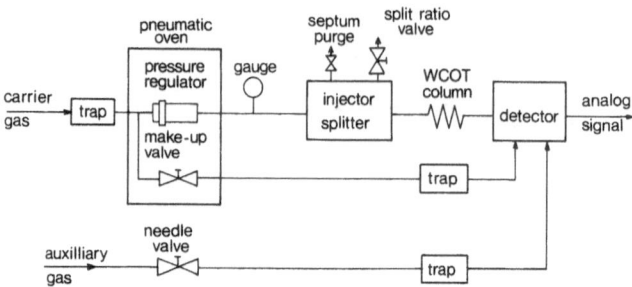

FIGURE 3.1. Pneumatic system and flow diagram for WCOT column operation.

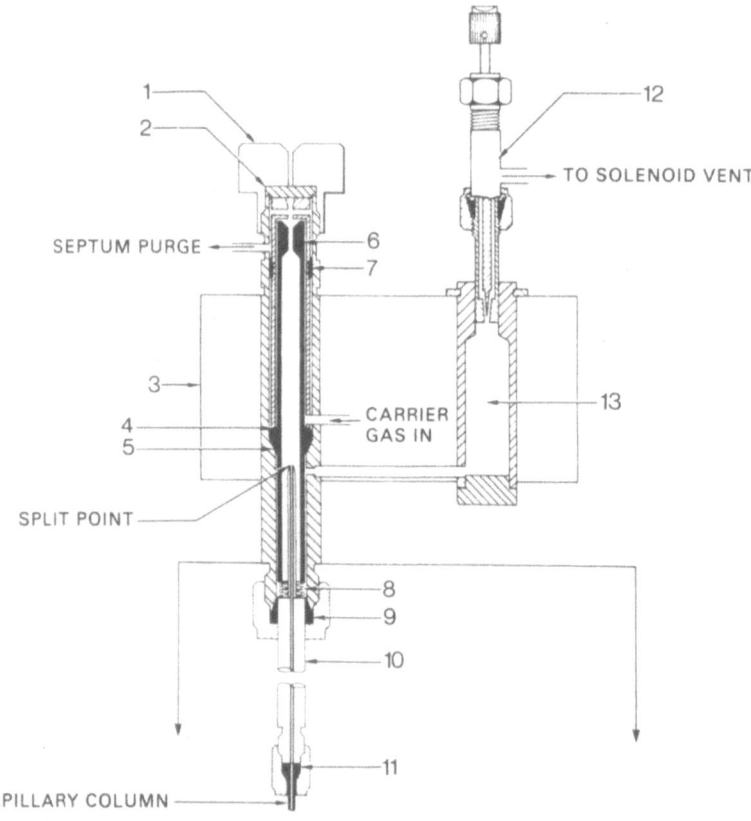

FIGURE 3.2. The split–splitless injector assembly (Varian Assoc., Inc.). (1) injector nut; (2) septum; (3) heated injector block; (4) washer; (5, 9, 11) graphite ferrule; (6) glass insert; (7) Teflon seal; (8) spring; (10) splitter; (12) split ratio valve; (13) buffer volume.

by a stainless steel body. The splitter part holds the glass insert, and the buffer part eliminates flow rate changes during splitting by ensuring that the sample has passed the splitting point before it reaches the split ratio valve. Positive septum purge is provided by venting septum bleed from back flush in the injector through a restrictor. This eliminates sorption onto the septum and prevents septum bleed from entering the column.

When the direct injector shown in Fig. 3.3 is employed, the split tip is replaced with a long, heavy-wall capillary glass insert. The glass insert is held in the same manner as the splitter but projects through the bottom of the injector–splitter into the oven.

As shown in Fig. 3.4, different glass vaporization tubes for homogenization of sample vapor with a carrier gas are available. The splitless-open, baffle, and frit vaporization tubes have short, fine capillary sections at the top to reduce back flush of the sample vapor. The open vaporization tube is employed for splitless injections. It may be packed with a quartz

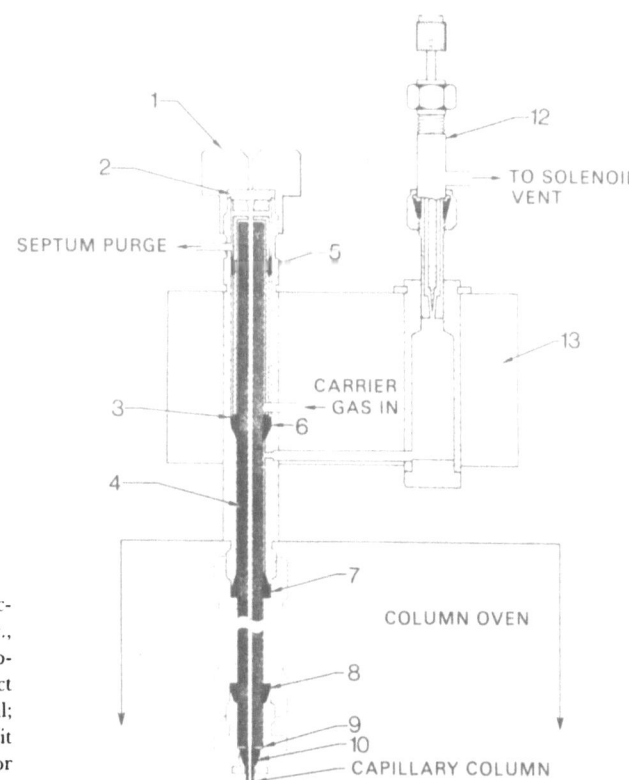

FIGURE 3.3. The direct injector assembly (Varian Assoc., Inc.). (1) injector nut; (2) septum; (3, 9) washer; (4) direct injector insert; (5) Teflon seal; (6, 7, 8, 10) ferrules; (12) split ratio valve; (13) heated injector block.

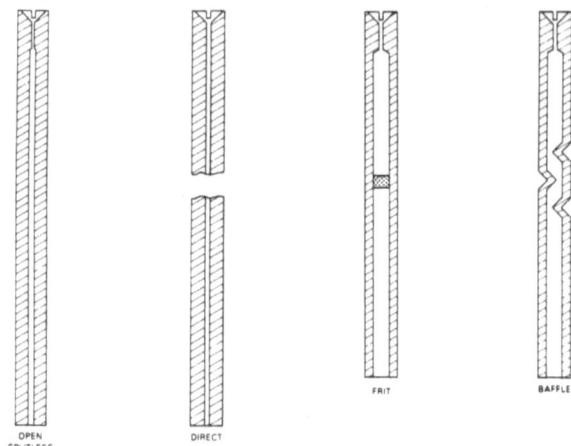

FIGURE 3.4. The capillary injector vaporization inserts (Varian Assoc., Inc.).

or glass wool. The baffle insert has less surface area and provides better heat transfer for sample vaporization and minimizes adsorption of polar components. The frit vaporization insert has a short sintered-glass frit fused to its wall to ensure linear splitting and to provide a high heat capacity for sample vaporization. It can be also filled with liquid phase of a suitable column-packing material to retain nonvolatile components from the sample. The fourth vaporization tube has a uniform capillary and is used as a direct injector primarily with wide bore WCOT columns and SCOT columns.

Packings with quartz wool or deactivated glass wool, which are tightly packed, have proven to be superior to other means of homogenization with regard to standard deviation of the repeatability and accuracy of relative peak area measurements. This insert improves homogenization of sample vapor carrier gas mixing and does minimize molecular weight discrimination. However, even silanized-glass wool may give rise to decomposition of sensitive sample components. In the split mode of sampling, very short residence times of the sample vapor in the splitter are acceptable only when the split ratio is high. Short residence time is optimal for a short inlet plug, and it minimizes the thermal and catalytic decomposition of the sensitive sample, but it requires complete homogenization of the sample carrier gas mixture.[2,3]

Quantitative evaluation of several vaporization inserts is shown in Fig. 3.5. Recently, however, investigation of the long and tight glass wool plug did not show that it improves significantly effectiveness of split injectors as related to the standard deviation.[4]

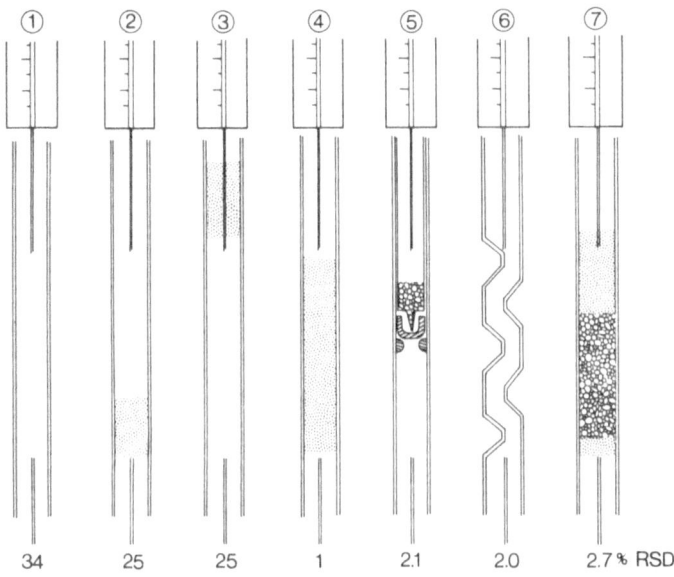

FIGURE 3.5. The quantitative evaluation of vaporization inserts. Relative standard deviations are given as percent RSD.[2]

Technical difficulties involved in sample introduction onto WCOT columns are more pronounced when quantitative analyses of trace components in complex mixtures are performed. The identification and quantitation depends upon:

(i)   an adequate separation efficiency,
(ii)   reproducibility and repeatability of retention data,
(iii) high integration and proper computation precision and accuracy on the peak area measurement, and
(iv)   no thermal and catalytic decomposition of thermally sensitive sample components on the column or in the system.

3.2. SAMPLE INTRODUCTION TECHNIQUES

A universal method for introducing a sample onto a WCOT or a packed column is by means of a microsyringe. It is important to realize that quantitative results depend on the quality of the syringe and the technique employed.

It should be noted that the critical parameter is the relatively long residence time of sample vapor in the injector. This may contribute to both decomposition and absorption of heat-sensitive substances since the contact time is sufficient to cause adsorption or condensation on relatively cold surfaces. Since the adsorption effect is directly related to the molecular weight, quantitative discrimination of higher molecular weight components will occur unless residence time is minimized. The general approach to reduce discrimination may be obtained by an uniformly heated injector, a long splitting time, a suitable solvent with a boiling point higher than the initial column temperature in order to achieve a solvent effect, and a slow injection. The purpose of the slow injection is to minimize the injector surface contact by the sample vapor. This may affect the discrimination of the high-boiling components in the mixture.

The poorest method for quantitative analysis is to retain the sample in the needle when introduced into the vaporizer tube.[9] The sample should be pulled back into the barrel of the syringe and the needle allowed to warm up to the injector temperature before the sample is transferred into it. Individual techniques of injecting a sample onto a WCOT column are described below.

3.2.1. Cold Syringe Needle Injection

The cold syringe needle injection technique is performed when the sample is taken back into the barrel of the syringe, the needle is introduced via septum, and the plunger pressed without delay to transfer the sample through the needle onto the WCOT column. There is no warming-up time for the needle to equilibrate with the injector temperature.

3.2.2. Filled Syringe Needle Injection

Filled syringe needle injection is the most often used technique, and it is obvious that this technique is used in laboratories where automatic injectors are employed. As shown in Fig. 3.6, this technique produces a significant discrimination for the high-boiling components in the mixture, since the percentage of a sample component left in the syringe needle increases with the boiling point increase.

3.2.3. Air Plug Method

The air plug technique uses air instead of pure solvent between the plunger and the sample, which helps to expel the sample mechanically and to avoid evaporation from the needle. However, it is found that a small amount of sample always remains in front of the plunger. This indicates that the air plug is not effective in moving the sample completely

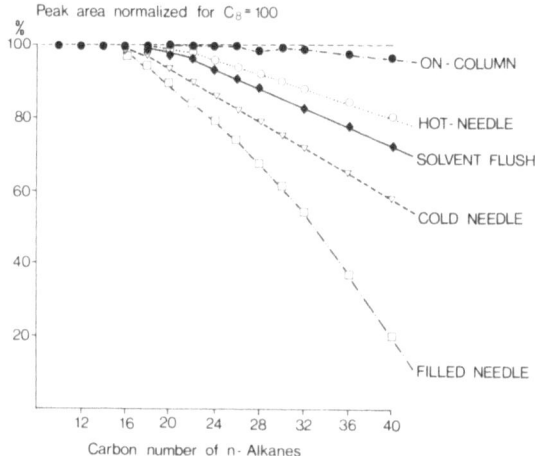

FIGURE 3.6. Discrimination of the *n*-alkanes obtained by different injection techniques.[2]

out from the needle. The rapid evaporation of the sample creates a back pressure that is sufficient to allow air to diffuse around the plunger and to sweep part of the sample backward toward the plunger (Fig. 3.7).

3.2.4. Hot Syringe Needle Injection

In contrast to injection with a cold needle, the empty needle is kept in the injector for up to 5 sec before the sample is collected in the syringe and introduced into the vaporization tube. During this time the needle

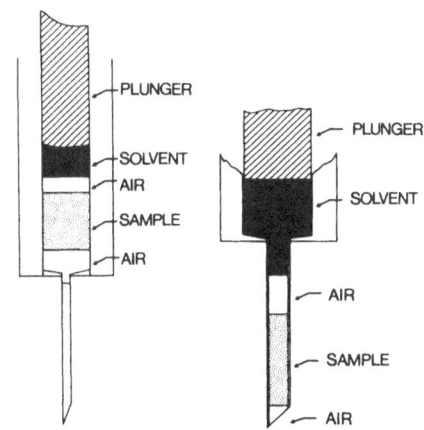

FIGURE 3.7. Air plug injection technique.

is warmed up to a temperature close to that of the injector temperature. As compared to the other splitless injection techniques, this one provides the best results. Apparently, instant evaporation of a relatively small portion of sample along the walls of the needle creates a high pressure which transfers the bulk of the sample into the vaporizer as a liquid. The hot syringe needle injection requires that the needle be preheated to the vaporizer tube temperature before the sample comes in contact with it, and transfer of the sample from the glass body of a syringe into the needle should be rapid in order to expel the liquid sample and to form a vapor "plug" on the rear side of the plunger, which will provide an optimum efficiency to transfer the entire sample into the vaporizer tube.

3.2.5. Solvent Flush Technique

The solvent flush method is based on the principle that the syringe is first filled with pure solvent.[10] The sample is placed in the syringe in such a way that it is located in front of the solvent and is separated from a solvent with an air space. The total content is then pulled back into the glass barrel of the syringe and injected rapidly using a cold or hot needle procedure.

It is possible to demonstrate the effect of different injection techniques by injecting a hydrocarbon mixture. Figure 3.8 indicates the result of discrimination for C_{10}–C_{24} hydrocarbons using "in-needle" and "hot

HOT NEEDLE INJECTION

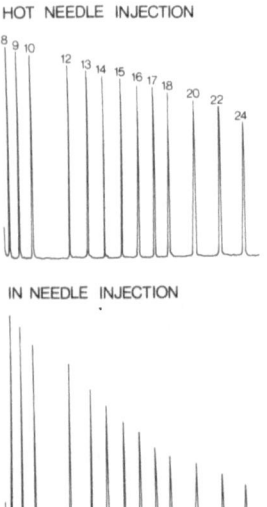

IN NEEDLE INJECTION

FIGURE 3.8. Chromatogram showing discrimination for C_{10}–C_{24} hydrocarbon mixture using hot needle and in-needle injection techniques.

TABLE 3.1
Characterization of Sampling Methods

Injection type	Concentration range	Type of column
Split sampling	More than 50 ng/component	Independent of column
Splitless	Less than 50 ng/component	WCOT columns
Direct	Less than 200 ng/component	WCOT and SCOT
On-column	Less than 100 ng/component	WCOT, PLOT, SCOT

needle" injections. A quantitative determination would be questionable in the case of "in-needle" injection.

Discrimination of components, especially those having much higher boiling points than those of the first portion in the chromatogram, is shown in Fig. 3.6 for different *n*-alkanes in the range of C-9 to C-44. The various sample introduction methods employed in WCOT gas chromatography and typical applications are shown in Table 3.1.[2] Each of these methods are discussed in some detail in subsequent sections.

3.3. SPLIT SAMPLING TECHNIQUE

The split sampling technique is considered unreliable with regard to precision and accuracy because of discrimination of sample components depending upon their polarity, molecular size, and concentrations in the mixture to be analyzed. Split sampling of complex mixtures may also be performed with an excellent qualitative and quantitative precision without being restricted to a limited range of parameters, such as sample volume, splitting ratio, and concentration range, by choosing a vaporization tube of suitable geometry and material and by proper heating of the insert, including the split region, the transfer lines, and the carrier gas preheating before entering the injector.

Split sampling can be employed with both isothermal and temperature-programming operations. Under optimal vaporization and homogenization conditions without aerosol formation in the glass liner, the adjusted injector temperature for samples with a wide range of volatilities is much higher than that of the WCOT column optimized for resolution. When the column is temperature programmed, the initial temperature may be so low that at least for the high-boiling components of the vaporized mixture, condensation or trapping in the first section of the WCOT column occurs. If the complete transfer of the whole sample is guaranteed, this trapping effect is desirable. The profile of the initial sample plug is narrowed, and the influence of the sampling procedure on the bandwidth of the separated components is minimized.

By employing the split sampling technique, the influence of the time of sample introduction and the transfer into the WCOT column on the final bandwidth is negligible because of the high carrier gas flow rates between the carrier gas inlet and splitting region. This technique is used for samples containing a range of trace to major component concentrations. It is particularly useful for trace or complex sample quantitation when solute components do not coelute with the solvent and when a specific detector is used where solvent peak will be suppressed.

The entire sample is injected directly to the injector system. This system incorporates sample injector, evaporator, and splitter into one unit and assures that the entire system including the vent restrictor, which controls the split ratio, is at one uniform temperature. The vent shut-off valve has a built-in leak, which maintains a very small carrier gas flow through the vent line even when the valve is closed in order to avoid any possibility of back diffusion. During sample introduction, the shut-off valve is open. It is advised that it be closed up to 60 sec after injection in order to maintain a constant carrier gas flow through the WCOT column.

Although the absolute value of the split ratio is usually unimportant for the evaluation of the analytical results, knowledge of its approximate value is advantageous when specifying the split restrictor and establishing the order of magnitude of the sample entering the column.

The split ratio can be conveniently established by measuring the respective flow rates, with the shut-off valve open, at the end of the column and the tubing coming out from the shut-off valve, and calculating their ratio. For example, if the column flow rate is 1 ml/min and the flow rate at the outlet of the shut-off valve is 150 ml/min, then the split ratio is 1 : 150.

3.4. SPLITLESS SAMPLING TECHNIQUE

In environmental and biochemical trace analysis the application of split sampling employing high splitting ratios is not acceptable due to the fact that it is necessary to introduce the entire sample onto a WCOT column in order to obtain adequate sensitivity. The sample load on the column for the trace components can be increased only by decreasing the splitting ratio, whereas the volume of sample injected can be increased only with respect to the volume of the vaporization liner. If the splitter flow rate is decreased to zero, all of the vaporized sample is introduced into the column. In this instance the column flow rate determines the rate of sample vapors transfer to the column inlet.

The splitless sampling technique allows solvent to be condensed at the head of the column after direct injection of the sample. The solute molecules are reconcentrated in the condensed solvent plug during the sampling process, and therefore large sample volumes can be injected with no significant injector-overloading effect. During the subsequent chromatographic process, the condensed solvent molecules act as a secondary stationary phase.

This phenomenon was first observed by Deans.[3] Later it was developed by Grob[4] and a theoretical explanation of this method described by Kaiser,[5] Schomburg,[2] and Grob.[6] The splitless injection technique, consists of introducing a relatively large amount of sample (1–8 μl) onto a WCOT column. During the injection step and the subsequent transfer of the vaporized sample into the column, the splitting valve remains closed. The components should be less than 50 ng per component or less (especially for fused-silica columns) to prevent overloading. Solutes eluting near the solvent partition rapidly in and out of the solvent liquid because of their solubility and similarity in chemical and physical properties. Solute reconcentrations and solute–solvent partitioning sharpen solute peaks eluting near the solvent because liquid–liquid diffusivity is small and solute peaks do not spread significantly during a substantial part of their elution time.

The theoretical explanation of the solvent effect was presented by Harris using Raoult's law and activity coefficients.[7] The vaporized sample is transferred onto the column essentially as a mixture. In the first stage of separation the solvent as the most volatile component in the mixture partitions in a liquid phase, at which point the front of the sample plug undergoes stronger retention than does the rear of the plug. The interaction of the solvent with the stationary liquid phase produces an area of very small phase ratio, β.[8] Any decrease in β causes a corresponding increase in the partition ratio, k'. The large k' values account for the reconcentration of solutes at the column inlet.

Guidelines for operating in the splitless injection mode with solvent effect follow:

1. The initial column oven temperature must be low so that solvent can condense at the head of the WCOT column. Generally, it is 20–40 °C below the boiling point of the solvent sufficient for achieving the solvent effect.
2. The concentration of the component being analyzed should not exceed 50 ng.

3. Sample volume between 1 and 10 µl can be used. The larger the sample volume, the better the solvent effect and the resolution between solvent and the solutes eluting near the solvent as shown in Fig. 3.9. However, the larger the sample volume, the faster the column degradation. Sample volumes between 1 and 2 µl are strongly recommended. A typical chromatographic temperature condition is 50–60 °C initial temperature, with a 2- to 3-min hold, and then temperature programming at 0.5–5 °C/ min to the final temperature with desired final temperature hold time.

4. It is important that the column inlet should be between 8 and 10 mm from the syringe needle to fulfil a basic principle of splitless injection, namely, to avoid unnecessary mixing of sample vapors with carrier gas during the sample transfer period. The vaporization of sample is produced next to the column inlet.

5. When volumes larger than 1 µl are injected, only the split valve should be closed, while the septum purge remains open maintaining a flow rate of 6–10 ml/min.

The injection is performed slowly to allow the sample vapor to enter the column without expanding excessively into the vaporizer tube. A minimum of mixing occurs since fresh vapor is produced continuously close to the column inlet. If injection is too rapid, sample vapor will be vented via the septum purge stream. This detail protects the injector parts and the carrier gas line from contamination. In this case it is possible to ensure a sufficient solvent effect even with injection times as long as 20 sec. Solvents such as n-hexane, isooctane, and cyclohexane produce excellent results. Methylene chloride, chloroform, and carbon disulfide may be used with a sample volume injection of less than 2 µl. Of course, chlorinated solvents should not be employed with ECD.

1µL128X   2µL128X   4µL512X   5µL512X

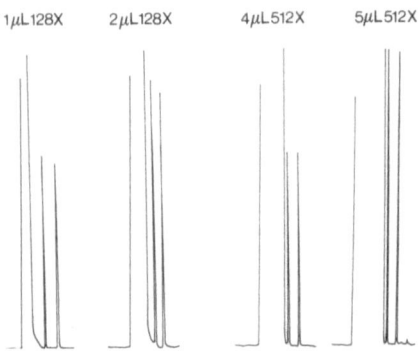

FIGURE 3.9. Chromatograms of benzene and toluene in n-hexane showing increasing solvent effect with sample volume. Injector temperature 60 °C; attenuations as shown.

The splitless injection with precolumn inlet configuration technique should be used when the sample contains nonvolatile and particulate matter. It is especially useful in handling environmental and biomedical samples. In this case up to 3 cm of 60–80 mesh solid support coated with Carbowax 20M (Aue packing) or SE-30 is packed in the injector insert between two silanized-quartz wool plugs. The sample must be injected into the precolumn bed to minimize initial peak broadening due to the backflashing. Peak broadening due to precolumn spreading must be reduced at the head of the WCOT column with a cold trap. The injector temperature must be high (more than 240 °C) so that the break-through time for the sample components in the precolumn is minimal.

3.5. ON-COLUMN INJECTION TECHNIQUES

A system in which a gold capillary tubing, permanently inserted into the 6-mm-i.d. injector body has been used.[11] An improved version of this WCOT column on-column injection system has been developed which includes both a micro- and macroversion of on-column direct sampling.[12,13] In the microversion the sample is placed in a small micropipette, and the carrier gas carries the sample onto the WCOT column. The macromethod is based on the principle that the sample is located in a small crucible which is inverted over the column inlet. Carrier gas forces the sample onto the column.

On-column injection techniques work perfectly well provided the temperature and flow conditions are fulfilled.[14] The sample should be evaporated from the capillary column wall to start the chromatographic process. The injector body is cooled to a maximum injector temperature of 40 °C if pentane or diethyl ether are to be used. The injection is located in the WCOT column fully within the temperature control of the oven, but not so distant that warming the injection needle with consequent evaporation from the needle occurs. The optimal distance depends on the design of the oven, but technically it is no problem to find optimal condition by utilizing a fused-silica needle syringe. Pressure and flow conditions play an important role. They are affected by the speed of introducing the sample onto the column, where sudden evaporation producing pressure shock would be undesirable and would result in a liquid plug migrating through the column. The plug will shorten and slowly disappear causing peak broadening. However, it also may produce a clogging of the column and dissolve the stationary phase.

In principle, on-column injection is governed by cold trapping of the sample or a solvent effect in order to concentrate the sample at the head of the WCOT column. The basic design of the on-column injector is shown in Fig. 3.10.

The injector has to fulfill two basic criteria: namely, that of guiding the syringe needle into the column and allowing septumless injection. Cooling is achieved by an airstream provided by the oven ventilator or compressed air tank. Recently, a modification has been applied to the Hamilton 32-gauge needle which may be replaced with a fused-silica needle.[27] The standard needle length is 75 mm. This puts the injection point 8 mm downstream (in Carlo Erba 4160 GC) from the cooled injector. In order to obtain exact quantitative reproducibility, small injection volumes and a lower column temperature are required avoiding partial evaporation from inside of the needle.

In another version, the on-column injector is a pressure-regulated system that utilizes a back-pressure regulator to control column pressure independently of the split flow.[16] A special alignment insert is used; the isolation valve replaces the septum; and a needle guide replaces the septum unit (Fig. 3.11).

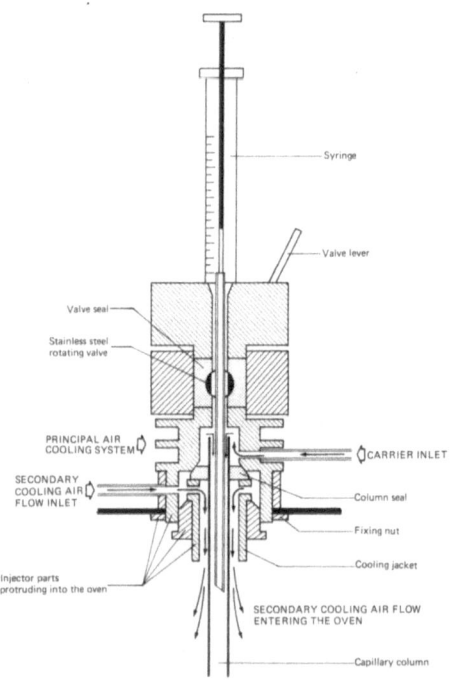

FIGURE 3.10. Cool on-column injector system. (Courtesy of Carlo Erba Strumentazione, Italy.)

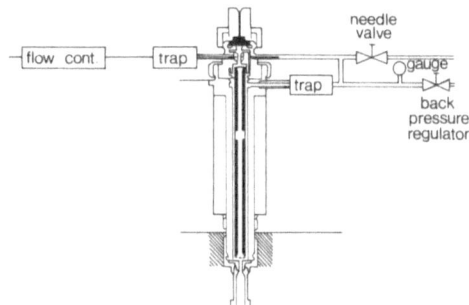

FIGURE 3.11. The Hewlett-Packard pressure-regulated cool on-column injector.

Since the column temperature is the determining factor of the system, the inlet temperature is always lower than the column temperature, thus no prevolatilization will occur as the injection is made. Also, the operating temperature of the isolation valve is between 30 and 40 °C.

When the inlet is air cooled, the cycle is reduced to a more acceptable shorter cycle time than when inlet system is insulated or uninsulated. Another reason for externally cooling the injector is to prevent the bottom portion of the inlet, which protrudes into the oven, from heating up. A third reason to cool the inlet during the injection is to prevent the vaporization of the solvent in those situations where the initial column temperature is well above the boiling point of the solvent.

On-column cool injection is useful when thermally labile compounds or high molecular weight compounds are being analyzed. Because the sample is placed directly on the column without a flash vaporization step, those substances that may decompose or rearrange during a conventional injection will be unaffected.

Figure 3.12 illustrates analyses of aliphatic fraction of south Louisiana crude oil under optimum conditions for the resolution of the pristane–phytane region. While the earlier sections appear heavily condensed, the last part is omitted.[15]

When a 0.5-μl solution is injected on-column at 120 °C, separation of more volatile components is poor (as shown in Fig. 3.12) due to the small solvent effect which is absent at 120 °C (b.p. *n*-hexane 69 °C). The solvent effect is pronounced when column temperature is lowered to 80 °C, as shown in Fig. 3.12. The principal merits of on-column injection may be summarized as follows:

1. Increased precision and accuracy of quantitative results and decreased discrimination caused by differences in volatility, concentration, or polarity of a sample is obtained. Because the sam-

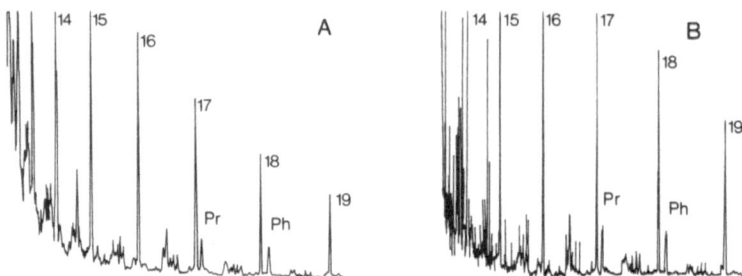

FIGURE 3.12. A portion of the chromatograms of aliphatic hydrocarbons of South Louisiana crude. (A) On-column-injection at 120 °C; (B) cool on-column injection at 80 °C.

ple is introduced at the column temperature, the eventual sample decomposition is minmized especially with thermally labile components.

2. If relatively clean samples, which do not contain nonvolatile components are analyzed, column deterioration is minimal. Also, no sample splitting occurs and quantitative recovery of high-boiling components is achieved.

3. Samples must be introduced directly onto the glass WCOT column. For this reason the entrance of the capillary should be properly shaped and its diameter (inner) must be at least 0.25-mm i.d. Syringe needle discrimination that represents one of the largest sources of error in quantitative analyses for a wide boiling point range samples is considerably reduced.

4. Parameters governing the solvent effect are column temperature, solvent volatility, and concentration of solvent vapor. Under the conditions of on-column injection, the concentration of solvent vapor at the point of injection, even with the smallest practical sample sizes, is always above the critical limit. Shown in Fig. 3.13 is the chromatogram of a solution of dichlorobenzenes in n-hexane containing broad range of injection volumes. Concentration per component ranging between 100 pg/μl to 10 ng/μl were injected, and injection time varied accordingly. A 25-m OV-17, 0.25-mm glass WCOT column was employed. Detection by means of an electron capture detector in the constant-frequency mode using hydrogen as a carrier gas at 2 and 40 ml/min nitrogen as an auxilliary gas was used.

The results indicate that the proper solvent must be chosen in order to produce narrow peaks. Under proper conditions, the solvent peak will not affect even closely eluting components. Provided that the chromat-

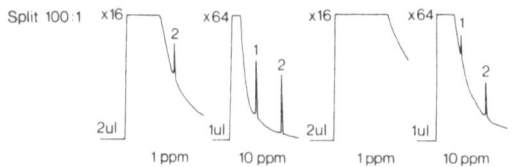

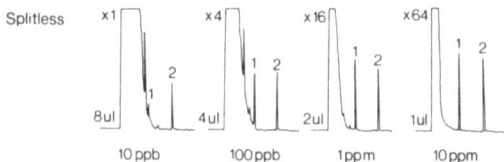

FIGURE 3.13. Chromatogram of dichlorobenzenes in *n*-hexane showing advantages of splitless versus split mode. (1) 1,3-dichlorobenzene; (2) 1,4-dichlorobenzene.

ographic parameters and conditions, such as solvent, stationary phase, and column temperature, are properly selected, no temperature difference is required and even isothermal analyses is possible.

In conclusion, the operation technique for the on-column injector is summarized in the following detailed steps:

1. The column is cooled down to a column temperature approximately 10 °C higher than the boiling point of the solvent used.
2. The syringe containing a measured amount of sample is inserted into the guide and is locked to the arrester (position 1).
3. The valve is open.
4. The syringe is pushed into the column (position 2).
5. The plunger is depressed slowly at a constant rate of 0.5 μl/sec.
6. The syringe is removed back to position 1.
7. The valve is closed.
8. The syringe is removed.

3.6. DIRECT INJECTION

The direct injection technique is based on introduction of the sample into the WCOT column without previous vaporization outside the column. The column flow in a direct injection mode can be either flow controlled or pressure regulated. Advantages of direct injection are related mostly to the volume, flow conditions, and surface properties of the injector, which operates in septumless and splitless mode. A syringe may even be replaced with a constant-volume pipette.

Disadvantages are related to the nonvolatile constituents present in the sample and their deposition in the inlet system. If larger volumes are injected, the stationary phase may be washed away from the surface and deterioration will occur with possible plugging of the orifice of the capillary column. Direct injection should not be confused with on-column injection technique.

The potential of this technique is not fully developed yet, and especially when high-performance WCOT gas chromatography in tandem with mass spectrometry is employed for analysis of high-boiling and extremely toxic components, such as 2,3,7,8-tetrachloro dibenzo-p-dioxin, significant advantages over all before-mentioned techniques are obtained.

The quantitative data for the temperature-programmed analysis of the n-alkanes C_{18} to C_{34} employing the direct injection sampling technique and having wide boiling point range between 196 to 492 °C is shown in Fig. 3.14.

3.7. COLUMN-SWITCHING TECHNIQUES

The use of column switching has been described relatively early for laboratory and process gas chromatography for a solution of various specific problems. Several review articles deal with this subject in great details.[17-19] A review of switching valves employing three separation columns and a slide valve has been described.[20] A method has been reported of a valveless column-switching technique where a sample does not come into contact with valves and valves are not thermally exposed.[21,22] This switching technique has been applied to the WCOT column as the multidimensional gas chromatographic method of analysis.[17] Multidimen-

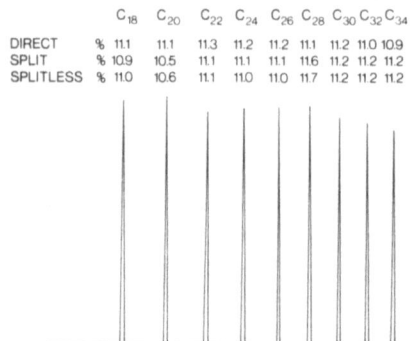

		C_{18}	C_{20}	C_{22}	C_{24}	C_{26}	C_{28}	C_{30}	C_{32}	C_{34}
DIRECT	%	11.1	11.1	11.3	11.2	11.2	11.1	11.2	11.0	10.9
SPLIT	%	10.9	10.5	11.1	11.1	11.1	11.6	11.2	11.2	11.2
SPLITLESS	%	11.0	10.6	11.1	11.0	11.0	11.7	11.2	11.2	11.2

FIGURE 3.14. Comparison of three injection techniques showing area percentages of individual C_{18} to C_{34} n-hydrocarbons.[2]

sional gas chromatography represents a separation technique whereby, during a single run, more than one chromatographic column is employed which may have similar or varying polarity or different lengths. Also, the directions of the carrier gas streams can be different in order to obtain preferential flows for separation of one or more selected groups of compounds eluting from the first chromatographic column.

In general, this method can result in a shorter analysis time and improve the qualitative and quantitative evaluation of complex mixtures with only a single injection. Further, it may prevent the deterioration of a column containing high boiling components or allow a more accurate determination of trace components which are observed by the tailing of a major peak. The two major areas of application are in the analysis of complex mixtures and the determination of trace components.

In the case of complex mixtures, groups of peaks that show poor resolution or which are not resolved at all on the first column may be selectively transferred for separation to a second column of different polarity. The flow-switching technique is essentially based on a pressure-balanced system. The switching is performed by magnetic solenoid valves set in a matrix that may be adjusted and automatically switched according to a preprogrammed timing mechanism (Fig. 3.15).

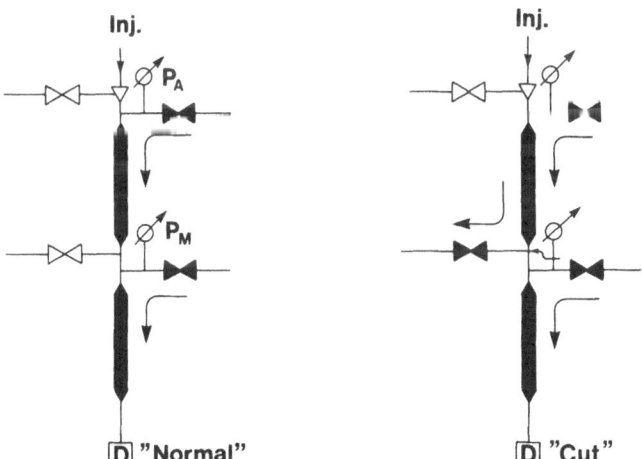

FIGURE 3.15. Dean's switching techniques. Normal two-column operation: P_A the inlet pressure corresponding to an optimum flow rate through both columns; P_M the inlet pressure corresponding to the appropriate flow rate for the second column; $P_A—P_M$ corresponds to the pressure drop of the first column which serves as a precolumn. Cut operation: indicates heart cutting on the first column; black valve indicates open valve; white indicates closed valve.

There are at least three commercially available systems on the market such as Siemens System L402—all-glass double-column system, Packard 429 Series gas chromatographs, and DANI MCF396 multicolumn flow control accessory.

The Dani MCF396 system consists of two modules (Fig. 3.16):

1. The operative module MFC396 installed as close as possible to the gas chromatographic cover consists of pressure regulators for carrier gas supply lines, micrometric valves for venting, and servovalves for line control. This module contains the circuit for the trap control by means of a Dewar cooler and a heater. Manual or automatic sequences are available.

2. The program module ACC 364 for the automatic timing of the selected values configuration allows 10 different sequences activation with a high degree of accuracy or selected valve configuration.

The trapping device is activated by the "cool-switch" valve, which regulates a nitrogen flow through a coil contained in a Dewar flask filled with a cryogenic medium. The coolant stream flows outside the trap,

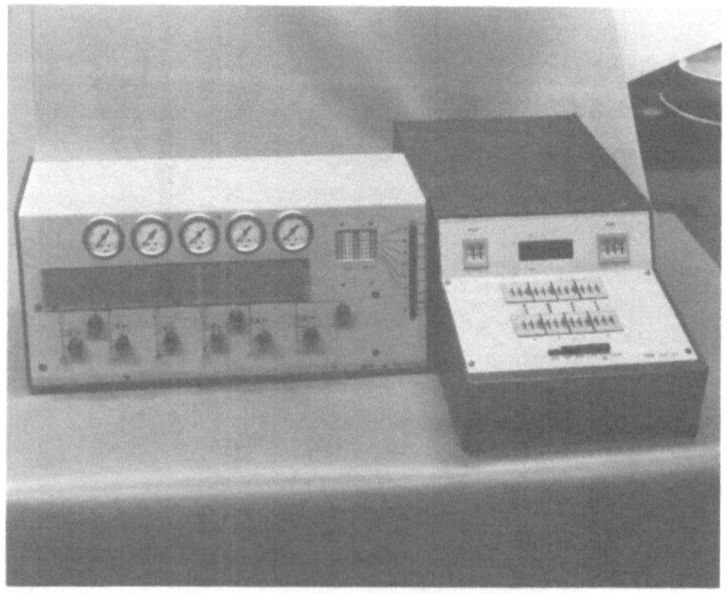

FIGURE 3.16. Column-switching system—DANI Model MCF 396—and its schematic.

leaving the trapped components within the trap. When the trapping cycle is over, the coolant, nitrogen, is shut off and replaced by a stream of preheated nitrogen which releases the trapped components. The various operative cycles of these valves are preprogrammed on the horizontal lines of the diode matrix and can step from one line to the next.

In order to facilitate the ease of operation in laying out a configuration, the module is provided with an interchangeable graphic panel where the flow diagrams can be displayed. Operating capabilities are described in the following examples.

3.7.1. Splitless Injection

Splitless injection requires the simplest valves configuration. Only two valves mounted on the venting lines of the top and bottom splitter are utilized. Just before the injection, valves 1 and 3 are closed and re-opened after a preselected time.

An advantage of the splitless injection and the septum wash may be seen in Fig. 3.14. Better separation is obtained by preventing the retardation of the sample in the solvent.

It is demonstrated that separation of benzene and toluene in n-hexane solution can be achieved at very diluted solutions using both the splitless injection and septum flashing.

3.7.2. The Back-Flushing Method

Back flushing is applied when groups of peaks, which are not of interest, are not wanted in the chromatogram, such as a group of high-boiling components. The components retained on the first part of the capillary or packed column can be vented to the atmosphere according to the scheme shown in Fig. 3.17.

At the injection the carrier gas flows through both chromatographic columns. At the back-flushing cycle valve 2 is closed and valve 4 is opened. Thus, carrier gas flow through the first separation column is reversed and the chromatography of lighter fractions continues through the second column.

In the following example (Fig. 3.18) a technical sample of perchloroethylene was injected onto a 5-m-long WCOT column coated with SP-1000. The first chromatogram shows a major part of the sample. The second chromatogram represents the sample after back flushing was introduced after 140 sec from injection. The major component was cut off completely. Conditions: 2×5 m SP-1000 glass capillary 0.23-mm i.d. Column temperature: 35 °C; injection 1 μl; detector FID.

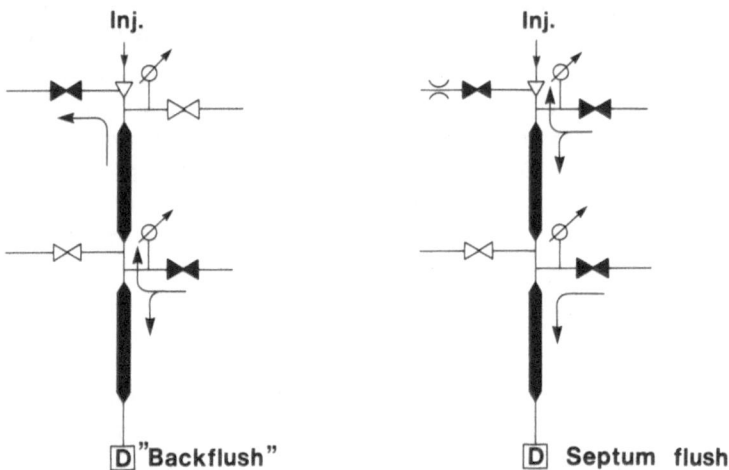

FIGURE 3.17. Diagram of back flush and septum flush (splitless) techniques.

3.7.3. Preseparation with Packed-Column and Intermediate-Peak Trapping

This configuration represents the most common way of preseparation and successive high-resolution determination with a WCOT column. When a packed column is not sufficient to give correct indications on the qualitative and quantitative component distribution in a multicom-

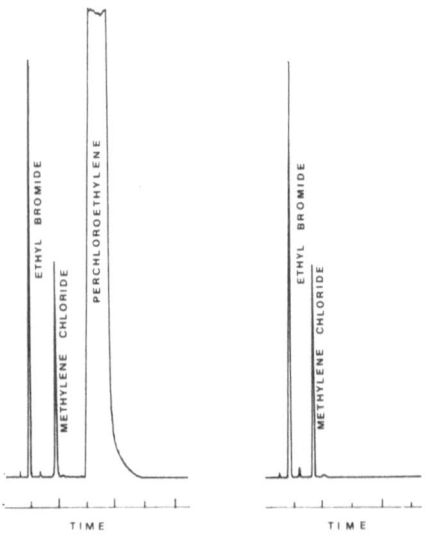

FIGURE 3.18. Chromatogram of impurities in perchloroethylene. Perchloroethylene peak was back flushed after methylene chloride elution.

ponent sample, high separation efficiency can be obtained using WCOT columns. However, WCOT columns lose their performance very quickly whenever large amounts of solvents are injected. Until recently, the split-less injection was quite suitable. The method, however, has the disadvantage in trace analysis of volatiles where traces of volatile components are removed almost in the same proportion as the solvent. The trapping method in combination with the heart cutting and back flushing offers the possibility to remove all the unwanted parts of the sample. The intermediate trapping may be highly useful for enrichment in trace analysis, as well as in eliminating the peak-broadening effect or distortion caused by a precolumn. Once the selected cut has been trapped, the precolumn may be back flushed and high-resolution separation continued in the WCOT column.

Figure 3.19 shows the analysis of trace amounts of pentachlorophenol.[23] Heart cutting was applied in order to remove the large amount of the major component by re-injecting the relevant part of the eluate from the preseparation. Venting off the eluate between the two columns is stopped at the point marked by a vertical dotted line in the chromatogram. The eluate flow is then directed into the main column, and the sample components contained in it are trapped in the first portion of the WCOT column by applying cold nitrogen to the trap. After the significant

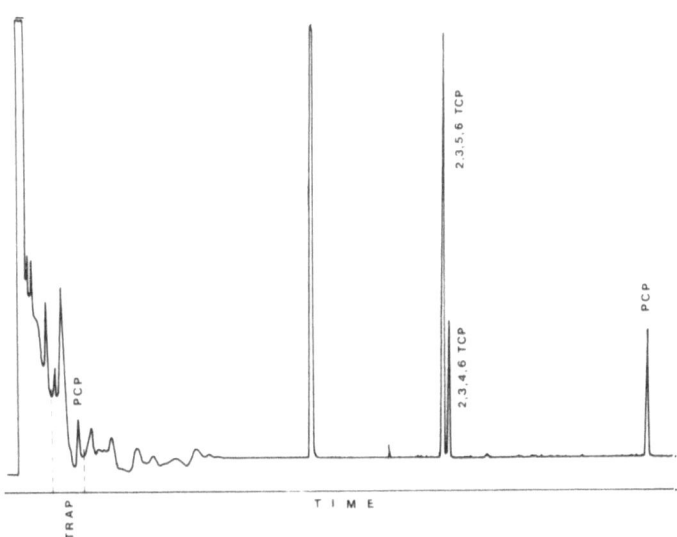

FIGURE 3.19. Heart-cutting technique applied to trap tetrachlorophenol and pentachlorophenol impurities.

parts of the eluate from the preseparation reached the WCOT column, the trap is heated and the trapped portion separated. In this case the column was coated with Carbowax 20M, 20 m in length, and run isothermally at 190 °C.

There are other more challenging possibilities that may be applied employing multidimensional chromatography.[17]

3.7.4. Preseparation with Packed-Column, Dual Injection, and Intermediate-Peak Trapping

This setup, shown in Fig. 3.20, allows heart-cutting operations to be performed in the precolumn. The effect is simplified using the right-side injector. The portion of the components of no interest may be discarded via valve 1, through the left vaporizer outlet. Then, by reversing the flow in the packed column (valve 3 closed), it is possible to trap the portion of interest and again discharge the rest (via valve 5, open). This operation may even be repeated to obtain an enrichment.

This technique may be employed when resolution of a packed column is insufficient for the separation of a suitable cut designated for subsequent separation with the high-resolution column. Then the preseparation can also be carried out with a WCOT column. The intermediate trapping of a desired cut may still be convenient, mainly for the determination of retention indices when using two columns of different polarity. The precolumn WCOT column may also be back flushed for the rapid elimination of the high-boiling components.

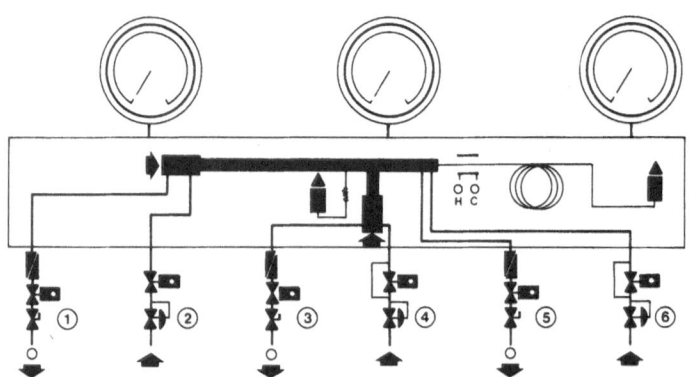

FIGURE 3.20. Preseparation technique employing packed-column–WCOT column combination, dual injection system, and intermediate-peak trapping.

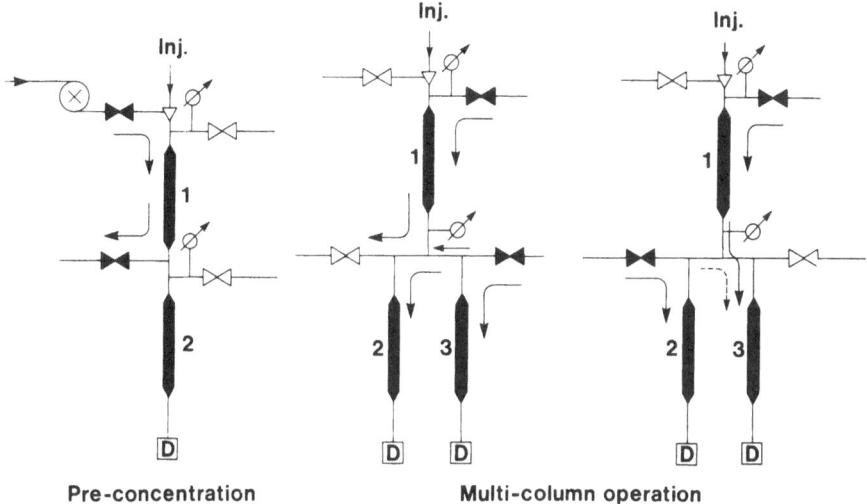

Pre-concentration   Multi-column operation

FIGURE 3.21. Schematics of various switching techniques employing two or three columns.

An elegant application of isothermal dual WCOT column chromatography for identification of selected components of a perfume oil has been reported.[17] Figure 3.21 shows applications of various switching techniques that could be employed for ultratrace analysis and very difficult separations.

3.8. FLOW PROGRAMMING IN WCOT COLUMN GAS CHROMATOGRAPHY

Flow programming represents the gas chromatographic process in which the carrier gas flow rate through a column is continuously increased during the analysis by increasing the column inlet pressure.

In the separation of multicomponent samples great differences in the capacity factors k' are observed. This problem results in the poor separation of components with small k' values and long retention times and broad bands of compounds with large k' values. Of course, this problem may be solved satisfactorily by using temperature programming;[24] however, it may happen that some components in the mixture are thermally unstable and decompose at higher column temperature or the stationary phase starts to bleed at higher temperatures.

In order to improve the analysis, it is possible to manipulate the flow rate of the carrier gas.[25] Basically, for a column to be used effectively

with a wide range of flow rates, the column must possess a broad minimum
when its height equivalent to a theoretical plate (HETP) is plotted as a
function of average carrier gas velocity. With packed columns the curve
exhibits a sharp minimum, and the working flow range is quite limited.
On the other hand, to obtain a consistent shortening of the retention
time, the flow rate must be increased at least by a factor of 10. Packed
columns obviously cannot sustain a 10-fold increase in inlet pressure.
WCOT columns having a low pressure drop and operating with optimal
flow rates in the range of a few ml/min owing to a broad minimum of
the HETP curve makes flow programming a complementary technique
to temperature programming. By the proper choice of the flow program,
it is possible to obtain the same or better chromatogram as obtained with
temperature programming (Fig. 3.22).

The operating range of flow programming in terms of retention time
shortening is roughly equivalent to a 100 °C of temperature change, while
maintaining the loss of column efficiency within reasonable limits.

From the point of view of the column performance, it may be noted
that under a large range of flow programming, the loss of efficiency in
a WCOT column is limited. During flow programming, the value of the
gas flow rate at the outlet of the column may conceivably be increased
by a factor of 100, whereas the value of the average gas flow rate is only
increased by about one order of magnitude. A change of this size can
be tolerated during a run. In this respect the best carrier gas will be
hydrogen or helium. In the majority of practical applications by operating
at low temperature, the partition coefficient α would be increased and
thereby counteract the loss of resolution due to the higher flow rate.
Depending on the temperature at which the flow programming is operated, to operate at constant α would improve or decrease the resolution
of two adjacent peaks.

The most used detectors in WCOT column gas chromatography are
flame ionization (FID), electron capture (ECD), and alkali flame ionization
detectors. They combine a dead volume compatible with the high-resolution WCOT column flow rates and a sensitivity high enough to detect
the very small amount of the analyte. These detectors respond not to
the concentration but rather to the total amount of the component per
unit time. Since this response does not change while changing the carrier
gas flow rate, one should expect relative response values identical to
those obtained under constant-flow rate conditions. This is true only for
small variations of the carrier gas flow rates because the sensitivity of
FID and AFD depends also on the ratio $He-H_2$, and the standing current
of ECD depends also on the gas flow crossing the ionization chamber.

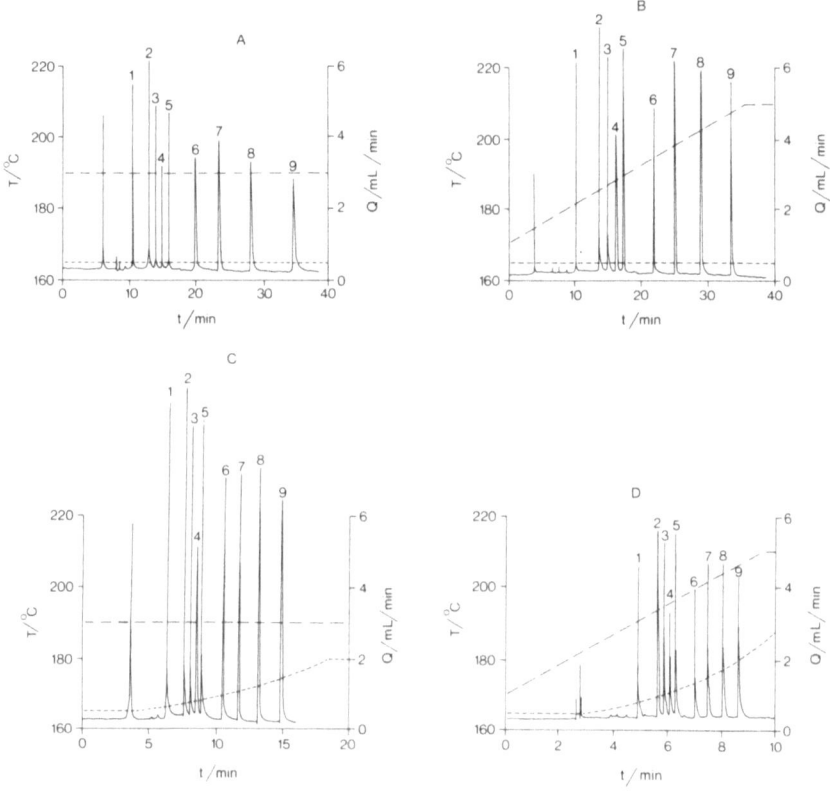

FIGURE 3.22. Flow programming and flow programming combined with temperature-programming analyses. (A) Isothermal–isobaric; (B) linear temperature programming, constant-flow system; (C) flow programming–constant-temperature system; (D) simultaneous flow and temperature programming system.

In the case of flow programming where the flow rate varies from one-tenth of ml/min up to several ml/min, it becomes compulsory to compensate the variation of flow rate entering the detector. The use of makeup gas should compensate for dead volumes and flow rate variations due to the change in column flow resistance when column temperature is programmed. By increasing the flow rate of the makeup gas to 30–50 ml/min, the difference in response during the flow programming is negligible both for FID and ECD. For the same reason the base line is steady even when operating at high sensitivity.[26]

REFERENCES

1. S. P. Cram, T. H. Risby, and L. R. Field, *Anal. Chem.* **52**, 324R–360R (1980).
2. G. Schomburg, H. Behlau, R. Dielmann, F. Weeke, and H. Husmann, *J. Chromatogr.* **142**, 87–102 (1977).
3. D. R. Deans, *Anal. Chem.* **43**, 2026 (1971).
4. K. Grob and K. Grob, Jr., *J. Chromatogr.* **94**, 53–64 (1974).
5. R. E. Kaiser, *Chromatographia* **9**, 337–352 (1976).
6. K. Grob and K. Grob, Jr., *J. HRC & CC* **1**, 57–63 (1978).
7. W. E. Harris, *J. Chromatogr. Sci.* **11**, 184–187 (1973).
8. W. G. Jennings, R. R. Freeman, and T. A. Rooney, *J. HRC & CC* **1**, 275–276 (1978).
9. K. Grob, Jr., and H. P. Neukom, *J. HRC & CC* **2**, 15–21 (1979).
10. E. C. Horning and M. G. Horning, in *Advances in Biochemical Engineering* (J. H. V. Brown and J. F. Dickson, eds.), Academic, London, 1972.
11. M. Verzele, M. Verstalpe, P. Sandra, E. Van Luchene, and A. Vuye, *J. Chromatogr. Sci.* **10**, 668–673 (1972).
12. K. Grob and K. Grob, Jr., *J. Chromatogr.* **151**, 311–320 (1978).
13. G. Schomburg and H. Husmann, *Chromatographia* **8**, 517–530 (1975).
14. K. Grob, *J. HRC & CC* **1**, 263–267 (1978).
15. F. Onuska, *Canadian Res. & Dev.* **12**(2), 26–33 (1979).
16. R. R. Freeman, K. B. Augenblick, and R. J. Phyllips, Hewlett-Packard, Tech., Paper No. 88, Avondale, PA, 1980.
17. G. Schomburg, H. Husmann, and F. Weeke, *J. Chromatogr.* **112**, 205–217 (1975).
18. W. Bertsch, *J. HRC & CC* **1**, 85–90 (1978).
19. W. Bertsch, *J. HRC & CC* **1**, 187–194 (1978).
20. W. Bertsch, *J. HRC & CC* **1**, 289–297 (1978).
21. H. Pauschmann, Tech. Lit. Perkin-Elmer, Bondensee, Federal Republic of Germany (1976).
22. D. R. Deans, *Chromatographia* **1**, 18 (1968).
23. F. I. Onuska, unpublished data.
24. S. Nygren, *J. Chromatogr.* **142**, 109–116 (1977).
25. S. Nygren and P. E. Mattsson, *J. Chromatogr.* **123**, 101–108 (1976).
26. F. Poy, DANI Technical Literature, Recent development in chromatography and electrophoresis, *Proceedings of the 9th International Symp., Riva del Garda, 1979*, Vol. 1, pp. 187–198, Elsevier, Amsterdam, 1979.
27. J & W High Resolution Chromatography Products, J & W Scientific, Inc., Davis, CA.

DETECTORS

Gas chromatographic detectors for environmental analysis must be sensitive to the minute amounts of contaminants being analyzed, but selective enough to discriminate against reasonable amounts of coexisting substrate materials. Despite this selectivity, it is necessary to protect the total gas chromatographic system by purifying extracts of the sample. This step will reduce the amount of impurities in the final solution to a level that will not be detrimental to the WCOT column or to the quality of the separation and measurement.

Nonselective detectors, such as the flame ionization detector (FID), produce complex chromatograms with peaks corresponding to all carbon-containing compounds in the sample.

4.1. FLAME IONIZATION DETECTOR

The FID combines the desirable characteristics of a linear dynamic range greater than 10^7, stability, good sensitivity ($\sim 10^{-10}$ g), and a general insensitivity to inorganic gases and water. Since the FID has a low effective volume, it is ideal for use with WCOT columns. The FID is a mass-sensitive detector. Its response is proportional to the total mass entering the detector per unit time, independent of concentration in the carrier gas. Originating almost 25 years ago, the FID has become the most widely used GC detector of all, retaining its original design and performance essentially unchanged.[1]

The usual FID uses a diffuse flame formed by the combustion of hydrogen and air-based oxygen to produce a low level of positive and negative ions in the carrier gas. The burner is insulated from ground and a polarizing potential (between 250 and 450 V) is applied to a narrow annulus around the jet tip and cylindrical collector electrode. Some detectors have parallel plates as electrodes in place of this cylindrical configuration.[2] When an organic compound is eluted from the GC, the quantity of ions formed increases significantly. The ion current collected is amplified and displayed on a chart recorder as the ions formed in the flame pass between the electrodes. The components of an FID are detailed in Fig. 4.1.

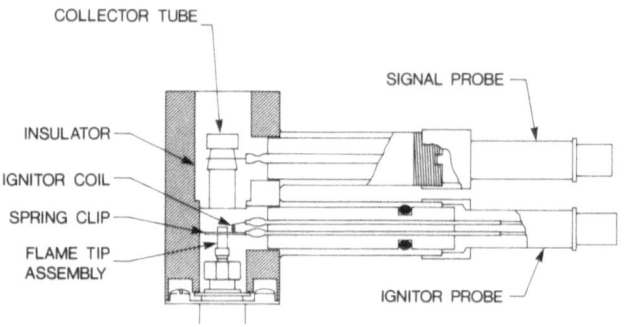

FIGURE 4.1. Flame ionization detector cell. (Courtesy of Varian Associates, Inc.)

The operating characteristics of the FID have been studied exten-sively.[3-5] Sternberg made a comprehensive attempt to understand the physical and chemical origins of its response, but with limited success.[6] Considering the complex set of free-radical and ion–molecule reactions occurring in the FID, it is not surprising that the basic mechanism is very complex in nature and that it is still the subject of continuous research.

Since the FID does not respond to the concentration of a component in the carrier gas entering it, but rather produces a signal which is pro-portional to the amount of organic material entering it per unit time, its sensitivity is generally expressed in the units of C/g or A/g sec. This value for the FID is of the order of 0.010–0.015 C/g.

The sensitivity of an FID is calculated from the following equation:

$$S = \frac{W_h \cdot h}{m} \tag{4.1}$$

where h is equal to the peak height (A), m is equal to the weight of sample introduced corresponding to the respective peak (g), and W_h is equal to the peak width at half height (sec).

4.1.1. Optimization of the FID

A flame ionization detector (FID) utilizes three different gases: the carrier gas, hydrogen, and air. There is an optimum hydrogen flow rate with each carrier gas flow rate at which the response and linearity of the detector are at a maximum. Therefore, it is important to optimize the hydrogen flow rate when working with the FID.

The situation is less critical with the air flow rate. The relationship of air flow to detector response is represented by a rising curve which levels off at approximately 280 ml/min. Therefore, it is only important to select a value within the plateau of the curve. Usually, an air flow of 350–600 ml/min lies in this range.

It is important to understand that when analyzing trace components present in a concentration below the mg/l or mg/kg level, the dynamic range has to be adjusted accordingly. For example, a dynamic range of 10^7 covers a range from 0.1 mg/kg to 100%. However, if the smallest peak represents a concentration of 0.01 mg/kg, then the same dynamic range extends only up to 1%. Therefore, caution is necessary if the calculation of very low and very high concentrations is to be carried out from a single chromatogram. A recommended practice is to use internal standards at different concentration levels so that any deviation from linearity can be compensated.

It is important to avoid detector overload in quantitative analysis. The FID is linear up to about 10^{-6} g/sec of an organic substance. It is advisable that when working with WCOT columns and a FID, the sample size never exceed 2 μl (liquid), with the exception that when trace impurities are analyzed, the peaks corresponding to high concentrations are not taken into consideration. If it exceeds the $1\cdot10^{-6}$ g/sec level, the sample size should be considerably reduced.

4.1.2. Detector Maintenance

Occasionally, the flame ionization detector may become contaminated. Generally, two types of contamination are common: liquid phase deposits and residues from samples.

The first type of contamination may occur if a WCOT column is heated to elevated temperatures with the carrier gas flow on but without the flame ignited. In such a case the liquid phase vapors might condense on the cold metal and ceramic surfaces of the detector. Therefore, during this step, it is always recommended that the WCOT column be disconnected from the detector. If such deposits are observed, the detector should be heated up to higher temperatures in order to evaporate the condensed substances with carrier gas flow on. If this treatment does not help, the detector has to be dismounted and cleaned with a solvent.

While the first type of contamination can be prevented by proper operation, the second type of contamination, that from sample residues, will always occur if either a silicon-type liquid phase is used or if the sample contains silyl derivatives, which are used frequently in the analysis of high-boiling substances such as sterols or phenols. Besides the usual volatile combustion products, these substances will always produce a fine white SiO_2 powder, which will be deposited on the jet and electrode. This deposit will result in a characteristic short time spiking noise.

It has been proposed that this fine SiO_2 deposit can be removed by injecting Freon 12 into the GC WCOT column. This will result in hydrogen fluoride formation among the combustion products, which in turn will react with SiO_2 forming volatile silicone fluorides. This method of detector cleaning may be used if the amount of SiO_2 deposit is limited, and it should certainly be the first attempt in detector cleaning. However, this treatment will not completely remove large deposits of SiO_2, which will have to be mechanically removed. This is done with a small flue brush.

For such cleaning, the collector electrode is pulled out and the detector cup removed. The brush is dipped into pure methanol and the jet, the collector electrode, and the contaminated parts of the detector are brushed gently until all the SiO_2 deposits are removed. The FID is then reassembled and, after the carrier gas flow is restored and the manifold is heated to above 120 °C, the flame is ignited and left on until the rest of the methanol is evaporated.

The flame ionization detector is a destructive type of detection system and is utilized for prescreening of samples prior to performing a gas chromatography/mass spectrometry (GC/MS) analysis in order to establish conditions and parameters necessary for optimum mass spectra.

4.2. ELECTRON CAPTURE DETECTOR

4.2.1. Principles and Operation of the ECD

From the time of its report in 1960 by Lovelock and Lipsky,[7] the electron capture detector (ECD) has enjoyed a steady growth in development and use. The wealth of theoretical concepts applied to the electron and ion–molecule reactions involved has led to many studies to understand the basic phenomenon. The simplicity of a device giving such high sensitivity and selectivity to environmental classes of compounds, such as halogenated pesticides, has encouraged development of detectors with an ever increasing reliability. The present status of the ECD is adequately described in recent reviews.[8-11]

The detector consists of a radioactive source that emits high-energy β particles capable of ionizing the carrier gas to produce secondary electrons. Three steps are involved in electron capture detection. These steps are generation of thermal energy electrons, capture of some of these electrons by an electrophilic compound, and collection and measurement of the unreacted electrons.

High-energy β rays generate electrons of low energy in nitrogen according to the reaction

$$N_2 \xrightarrow{\beta} N_2^+ + e^- + N_2^*$$

Each β particle may generate 100–1000 thermal electrons before it has reduced its kinetic energy to thermal levels. The thermal electrons have mean energies of about $5 \cdot 10^{-2}$ eV compared to 67 and 18 keV for the β particles of ^{63}Ni and ^{3}H, respectively.

The thermal electrons next interact with the effluent from the chromatographic column. When an electrophilic compound, AB, enters the detector, electrons are removed via an electron-capturing process.

$$AB + e^- \rightleftharpoons AB^- \qquad \text{(nondisassociative capture)}$$

$$AB + e^- \longrightarrow A + B^- \qquad \text{(disassociative capture)}$$

The remaining electrons are then collected and the resulting current is measured.

In principle, the ECD is selective for highly electronegative compounds, but in practice it is the least selective of the widely used environmental detectors. Rigorous cleanup of environmental extracts is required to eliminate extraneous peaks due to compounds containing halogen, sulfur, nitrogen dioxide, phosphorus, and some PAH hydrocarbons. Its sensitivity, however, is the highest of any contemporary detector.

Many radiation sources have been investigated for use in the ECD. The most extensively used are ^{63}Ni and ^{3}H emitters. The ^{63}Ni cell offers many advantages as a radiation source because of its higher operating temperatures with less contamination problems. Some characteristics of radiation sources are summarized in Table 4.1.

The ^{63}Ni detector can be used safely to 400 °C without appreciable loss of radioactive material. This high operating temperature reduces the possibility of contamination from extract impurities or from bleeding of liquid phases. It also extends the number of compounds that can be detected and greatly reduces detector maintenance. The electron capture detector is employed with either a constant negative dc voltage or an intermittently pulsed voltage of constant frequency imposed across the anode–cathode. This pulsed mode improves performance over dc operation, and it is achieved by applying the detector voltage as a sequence

TABLE 4.1

Characteristics of Radiation Sources

Isotopes	^{63}Ni	Sc^3H	T^3H
Type of decay	β	β	β
Half-life in years	125	12.5	12.5
Maximum temperature (°C)	400	325	220
Standing current (A)	$1 \cdot 10^{-9}$	$1.3 \cdot 10^{-8}$	$1.3 \cdot 10^{-8}$
Specificity factor	10^7	10^7	10^7
Linearity range	50	800	500

of narrow pulses with a duration and amplitude sufficient to collect the very mobile electrons but not the heavier, slower negative ions. Figure 4.2 shows two pulsed modes of ECD operation.

The constant-current mode can be performed with frequency-modulated or variable-frequency modes of operation. A standing current is again achieved by applying voltage pulses, but in this case the pulse-sampling frequency is varied by a servomechanism closed-loop control circuit that maintains the standing current constant even when an electron-absorbing component enters the detector. The pulse frequency is converted to a dc signal, which is monitored in the usual way to provide a chromatographic trace. This mode of operation provides a high degree of baseline stability and an increased linear response range without loss of detectability.

The linearized ^{63}Ni constant-current variable-frequency mode of operation was introduced by Maggs.[12] In principle, when capturing species

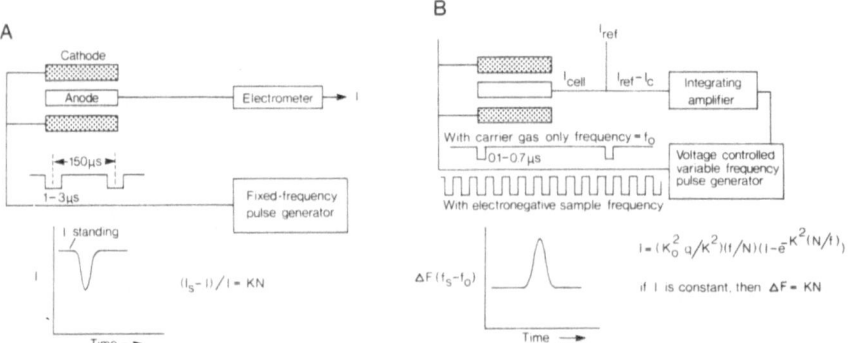

FIGURE 4.2. Two pulsed modes of ECD operation: (A) constant-frequency mode; (B) variable-frequency mode. I is electron current collected; I_s is electron current with only carrier gas; $K_1 K_0$ = cell constants; N = sample quantity.[2]

enter the cell, electrons are removed and the number collected with each voltage pulse is reduced. In order to maintain a constant electron current, the voltage must be pulsed more frequently.

The pulsed, linearized ECD allows detection of low picogram amounts of chlorinated and halogenated compounds using isothermal or temperature-programmed operation and gives a linear dynamic range of 10^5. This compound-independent, extended linearity is of great benefit for automated analyses where a wide concentration range of samples can be analyzed without readjustment of a sample volume. An ECD with a small cell volume of about 0.3 ml is well suited to WCOT column instruments. Most pulsed electron capture cells permit efficient electron collection employing nitrogen carrier gas.[13]

High-temperature ^{63}Ni detectors are probably in greatest general use, followed by the concentric-design tritium detectors. A benefit of using ^{63}Ni foil as the radioactive source is that the detector may be operated at elevated temperatures. This reduces detector contamination by column bleed and by less volatile sample components. Should contamination occur, it is frequently possible to clean the nickel foil by injecting 100 μl of water a few times into a detector heated to 300 °C using an empty column. Purging the ECD at 400 °C overnight may also be helpful. Response of the ECD depends upon temperature, carrier flow rate, electrode and cell configuration, electrode position, amount of radioactivity, and contact potentials caused by adsorption of sample components on electrode surfaces and applied potential. The unpredictable nature of these parameters causes anomalous responses, drifting baseline, and variable sensitivity. Operating parameters must be optimized for the detector of each manufacturer.

It should be realized that the detector response to a given compound is temperature dependent, and an increase in sensitivity is sometimes found at a lower detector temperature. When a specific analysis is optimized, the temperature dependence of the ECD response should be characterized.

Because the ECD is an extremely sensitive detector, solvents used in GC techniques must be very pure. Each new lot of solvent should be tested to verify that there are no interfering impurities present. If a concentration step is required, the solvent residue blank should be monitored. Chlorinated solvents are inappropriate, and excess halogenated derivatizing reagent must be removed before the sample is injected. Water is generally not recommended as a sample solvent for analysis by a ECD, but it can be used if the column is water compatible.

Temperature programming can be used with WCOT columns and an ECD. Although detector response may change when the WCOT column is at different temperature due to variations in column bleed, these changes are reproducible and quantitative data are obtained. Normal problems associated with temperature programming, such as septum bleed and carrier gas impurities, tend to be magnified by the ECD because of its high sensitivity. Problems with septum bleed can be reduced by using a septum purge device and preconditioned septa.

4.2.2. Contamination of the ECD

Contamination of the ECD by deposition of a liquid phase on the electrodes seriously affects detector performance. This decreases a portion of the detector capability, leading to loss of sensitivity, trailing peaks, or erratic baselines. Sources of contamination may be a bleeding column, contaminated carrier gas, bleeding septum, contaminated sample inlet, dirty carrier gas flow controller, or contaminated products.

Tight connections of the ECD housing and the entire system are very important. Oxygen is a frequent contaminate in nitrogen carrier gas, and the ECD responds well to traces of oxygen. Usually, a background profile should be made after changing the tank of carrier gas and at least 1 hr allowed for the system to equilibrate. A suitable oxygen trap and a clean chromatographic system are most important prerequisites for good performance.[14]

Certain liquid phases tend to bleed in varying degrees at normal operating conditions, even after conditioning for extended periods of time. The high-temperature silicones (OV-, SE-, and SP-type gums) produce very low-bleed WCOT columns. Also solvents, monomers, and plasticizers can bleed from the septum and be swept through the WCOT column into the detector. Glass inlet lines used for splitless and on-column injection should be changed frequently.

Contamination from dirty tubing or other system components prior to the inlet can be caused by a tank of carrier gas containing oils, grease, or water. Use of a molecular sieve filter will usually prevent this problem. These adsorbent traps must be exchanged regularly. If contamination has accumulated in the tubing and flow controllers from the carrier gas, simply changing the tank may not solve the problem. The entire system would have to be flushed out with a low-boiling solvent.

4.2.3. Response Factors

ECD response factors are critically dependent on several instrumental parameters such as detector temperature, column temperature, column bleed, and carrier flow rate. All operating conditions must be controlled stringently to obtain day-to-day reproducibility and meaningful

TABLE 4.2
Relative Responses of ECD for Various Functional Groups

RRF	Functionality
10^0	Hydrocarbons like pentane, isooctane, benzene
10^1	Esters, ethers
10^2	Monochloro, monofluoro, alipathic alcohols, ketones
10^3	Dichloro-, difluoro-, monobromo- derivatives
10^4	Trichloro-, anhydrides
10^5	Monoiodo-, dibromo-, nitro derivatives
10^6	Diiodo-, tribromo-, polychlorinated aromatics

results. Selectivity and relative response of various functional groups to the electron capture detector are summarized in Table 4.2. Such a range of response gives the ECD considerable selectivity. Experimental values are shown in Table 4.3.

In the analysis of multicomponent mixtures where the response factor varies significantly, it may be almost impossible to find a suitable internal standard. However, it is necessary for a quantitative determination to employ one in order to obtain meaningful data.

4.2.4. Troubleshooting Electron Capture Detectors

Basically three types of problems occur when using ECD: low sensitivity, noise, and drifting baseline.

Low Sensitivity

1. Low sensitivity may be caused due to the ECD overload. This can occur even at about a 1-ng load. The peak exhibits a characteristic cigar-shaped apex. It may be diagnosed easily by reducing sample amount to one-half of the injected volume. If the new area is greater than calculated, diagnosing overload is correct. In this case a lower pulse rate, including additional purge flow, should be considered as well.

TABLE 4.3
Relative Response Factors Examples

Compound	RRF
1-chlorobutane	1
1-bromobutane	300
1-iodobutane	90,000
Tetrachloromethane	400,000

2. High purge rates will decrease sensitivity.
3. If there is no pulse, sensitivity will be very low. It is recommended to check this with an oscilloscope after consulting the circuit diagram. If the pulsing rate is switched from a lower setting to a higher one, the recorder pen should move rapidly upward. If it drifts slowly back after a rate change, cell contamination is likely.
4. A leak at the septum or WCOT column to injector fitting may cause low sensitivity. An increased retention time may be observed.

Noise

1. If the recorder is functioning normally, a noise-free baseline must be obtained with the attenuation at ∞.
2. A high flow rate through the ECD can result in noise. A carrier flow rate of about 2–3 ml/min and 60–80 ml/min purge flow rate for Ar–CH_4 is recommended.
3. Extraneous chemicals such as phthalates from septum bleed or column bleed and gas impurities can cause noise problems. These possibilities may be tested by cooling the column to a lower temperature, replacing septa with one of a low bleed which is well conditioned.

Drifting Baseline

1. A change in operating conditions such as detector temperature and a high temperature-programming rate will result in drift. Also, a change in the system such as a new septum or a new ferrule, carrier gas or purge gas supply will result in drift. An equilibration period of up to 24 hr may be necessary on restarting the instrument after a down period to reach full operating conditions.
2. If the system is overloaded with sample, or a halogented solvent has been injected, drift may last for several hours. This problem can be minimized by raising column flow rate and temperature of the WCOT column.
3. The most rapid way to check for a leak which can cause drift is to further tighten all connections inside the oven and ensure that the septum is properly installed.

4.2.5. Cleaning Electron Capture Detectors

When an ECD is dirty, the usual symptoms are loss of sensitivity, drift, or a negative peak on the peak tailing edge. Deposits can often be removed by using one of several cleaning steps.

4.2.5a. Solvent Washing. In this case the cold cell must be totally filled with a suitable solvent for at least 15 min. First methanol should be applied. After 15 min, it should be drained and filled with acetone, following with benzene or toluene for 15 min and finally with hexane. The last solvent should be removed by drying for 1 hr in an oven that has been preheated to 120 °C.

4.2.5b. Thermal Conditioning. Volatile deposits can be removed by raising the ECD temperature approximately 50 °C above the normal operating temperature, using normal flow rates and leaving the detector overnight or over a weekend.

It is frequently possible to decontaminate a ^{63}Ni foil in the ECD by injecting 100 μl of water a few times into a 350 °C system employing an empty column. Purging the detector at 400 °C overnight with an empty column may be also helpful.

4.3. THERMIONIC (ALKALI FLAME IONIZATION) DETECTOR

The thermionic emission detector (TED) (Fig. 4.3), also known as the alkali flame ionization detector (AFID), provides a response that is highly selective for compounds that contain phosphorus or nitrogen. The ionization source in a TED operates by thermally heating an alkali salt pellet or alkali glass bead source. Its sensitivity to boron and arsenic response has also been noted. The selectivity ratios for detection of

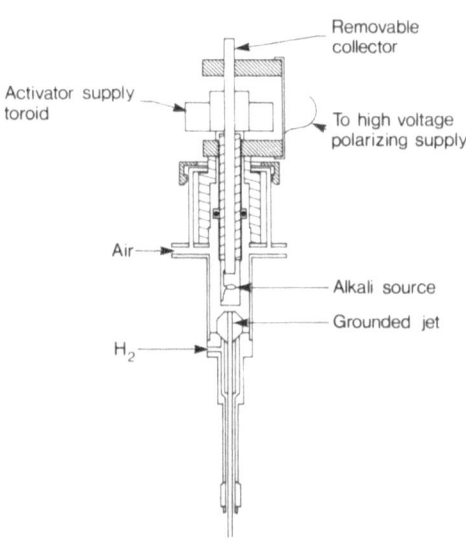

FIGURE 4.3. Thermionic detector cell. (Courtesy of Varian Associates, Inc.)

phosphorus- and nitrogen-containing compounds versus carbon detection are 75,000:1 and 35,000:1.

The first type of ionization source (AFID) operates by thermally heating an alkali metal salt in the presence of vapors containing nitrogen or phosphorus atoms. The emission of positive ions from heated metallic surfaces, first observed by Richardson in 1904, was thought to be due to alkali metal impurities in the heated metal.[15] Almost 60 years later Karmen and Guiffrida described the first TED for nitrogen- and phosphorus-containing compounds.[16] Since then, many detectors based on this principle have found wide application in the determination of sulfur, nitrogen, phosphorus, halogens, arsenic, and lead-containing compounds in environmental and pesticide analyses.

The design of these detectors has centered on construction modifications of the conventional FID. In contrast to the FID, the TED has been plagued with instability, which has made it unsuitable for routine analysis. A major source of instability lies in the heating method used earlier to produce thermionic emission from the alkali metal salt. In most cases the flame has been used for both hydrocarbon combustion and emission source heating. In 1966 Abel suggested heating the alkali source to improve stability.[17] This independent heating has been the key to the increased stability and reliability found in the TED used today.[18-20]

4.3.1. Mechanism of Selectivity of a TED Detector

The mechanism of response of the thermionic emission detector for nitrogen and phosphorus containing compounds (TED N–P) is not fully understood. However, a brief probable explanation of the selective detection can be deduced. The normal FID ion collection assembly is replaced with a collector that has a ceramic cylinder coated with an alkali salt activator as depicted in Fig. 4.3. The cylinder is geometrically centered above the flame jet and is heated electrically to a dull red color with a stepdown transformer. A negative-polarizing potential of 240 V is applied to the collector for species collection. The ion current generated by the thermionic emission in the presence of nitrogen- and phosphorus-containing compounds is measured with the conventional FID electrometer. The flame in this detector is not ignited, but rather is adjusted to give a low-temperature plasma. This plasma is used only for ion production and dissociation as the organic compounds elute from the column. The plasma is not used to heat the collector. Rather, a partial pyrolysis takes place, producing intermediate stable CN radicals from nitrogen-containing compounds. The radicals take electrons from the alkali, resulting in a cyanide ion, and a positive alkali ion migrates to the collector electrode and again liberates an electron. Collection of electrons creates the specific response. A similar mechanism has been proposed for phosphorus, except that PO or PO_2 are assumed to be the intermediate radicals.

Proof of this mechanism has been offered in the fact that most detectors do not respond to compounds in which HCN-type bonding does not exist. However, this mechanism recently has lost some credibility because some designs can respond to nitrogen compounds not containing HCN bonding.[21] It should be emphasized that there is no mode of the TEC detector selective for nitrogen-containing compounds only; there is only a strong response to phosphorus and nitrogen in the N–P mode.

4.3.2. Phosphorus Mode

The phosphorus mode is sensitive to phosphorus-containing compounds. A hot flame exists because of an increased hydrogen flow rate, and the jet of the detector is grounded. The compounds eluted from the column are fully burned, and the electrons produced by the normal combustion process are conducted to ground. The combustion products of phosphorus-containing compounds react with the alkali on the surface of the bead or ceramic cylinder and produce ions that are captured by the collector electrode, thus producing the response. Nitrogen-containing compounds give a reduced response in this mode of operation.

4.3.3. Selectivity and Response of TED

The TED is about 50 times more sensitive for nitrogen and 500 times more sensitive for phosphorus compared with the FID. The limit of detection for malathion is $6 \cdot 10^{-14}$ g/sec, or calculated for phosphorus $6 \cdot 10^{-15}$ g/sec. Nanogram to picogram quantities of most nitrogen-containing compounds can be determined. Sensitivity of the detector depends on detector background. The background is affected by bead heating current, gas flow rates, and quality of the bead. Operation of the detector at a fixed background current requires only occasional adjustment of the detector and has resulted in very uniform response time. The hydrogen, air, and helium or nitrogen carrier gas flow rates should be optimized for maximum signal-to-noise ratio for nitrogen- and phosphorus-containing compounds.

The thermionic detector is not usable with columns of liquid phases containing halogen, phosphorus, or nitrogen such as OV-210, XE-60. WCOT columns coated with Carbowax 20M, OV-1, SE-30, OV-17, and Apiezon L are recommended. H_3PO_4-treated supports, glass wool in the injector, and use of the Snoop-Leak detector should be avoided.

Care should be taken to turn off the collector voltage while changing the septum, WCOT columns, or gas cylinders. Cleaning with Freon will adversely affect the alkali salt or bead of the detector.

In summary, TED for detecting nitrogen- and phosphorus-containing compounds is generally the simplest, most sensitive, and least selective detector suitable for work with WCOT columns. Its optimum, reproducible use is more of an art than science. Reviews of the TED have been published.[22]

4.4. PHOTOIONIZATION DETECTOR

The photoionization detector (PID) possesses characteristics equivalent to or better than those of the flame ionization detector. Having a higher sensitivity and suitable linear dynamic range, it is a nondestructive detector and responds as a concentration-sensitive detector.[23] The PID appears to have many of the requirements necessary for WCOT columns. It consists of two modules, the detector and the power supply interconnected by a multiconductor cable. The detector consists of a sealed uv lamp that emits the Lyman α line of hydrogen at 10.2 eV through a magnesium fluoride window into the ionization chamber. The process of photoionization is initiated by absorption of a 10.2-eV photon by a molecule.

$$RH + h\nu = RH^+ + e^-$$

where RH^+ is an ionizable molecule and $h\nu$ is a photon with an energy larger than the ionization potential of the molecule. Figure 4.4 shows the principle of the photoionization detector.

This detector requires some modifications to meet the demands of a high-resolution WCOT column system.[23] The inlet tube to the detector is modified to allow the capillary column to be extended up into the detector and bring the column as close to the detector cell as possible.

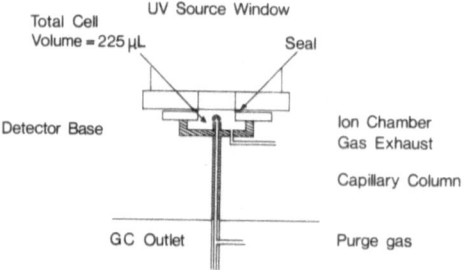

FIGURE 4.4. Photoionization detector cell.

TABLE 4.4
Summary of the Photoionization Detector Molar Responses[a]

Sensitivity increases as number of carbon atoms increases.

The molar sensitivity (S) for *n*-alkanes relative to benzene can be expressed as

$$S = 0.0715n - 0.457$$

where *n* is the carbon number.

Sensitivity for alkanes < alkenes < aromatics.

Sensitivity of alkanes < alcohols < esters < aldehydes < ketones.

Sensitivity of cyclic compounds > nonbranched compounds.

Sensitivity of fluorine-containing compounds < chlorine-containing ones < bromine-containing < iodine-containing compounds.

For substituted benzenes; electron-releasing groups increase sensitivity and electron-withdrawing groups decrease sensitivity.

These general conclusions should give an analyst of this detector a better understanding of its performance and applicability in environmental trace analysis.

---

[a] From Ref. 25.

However, this modification is not mandatory for fused-silica columns. The PID coupled with a WCOT column system has some advantages, namely, improved sensitivity and virtually no solvent front. The PID has been shown to respond to carbon-containing compounds in a manner similar to a FID; i.e., it is a carbon counter. The PID should be even more useful for trace analysis than the FID because of its 30-fold increase in sensitivity. Since the dynamic range of the photoionization detector is greater than 10^7, it will also be useful at lower levels. The nondestructive aspect of the PID should extend its usefulness when coupled in series with element-sensitive detectors for sulfur, phosphorus, and nitrogen.[24] Although the photoionization detector complements the FID in many respects, the PID should be considered as a sensitive detector in its own right, which has response to many organic and some inorganic compounds. The detector response relative to benzene on a molar basis for a large number of organic compounds has been reported.[25] Conclusions are summarized in Table 4.4.

4.5. THE MASS SPECTROMETER AS DETECTOR

In its simplest form the mass spectrometer performs three basic functions as a GC detector. These functions are to deliver a sample into the ion source at a pressure of 10^{-5} torr to produce ions from neutral molecules, and to separate and record a spectrum of ions according to their mass-to-charge ratios (m/z) and relative abundance.

If a WCOT column represents the heart of the system, the mass spectrometric detector is the brain at the highest intelligence level. It is capable of providing both qualitative and quantitative data by means of spectral interpretation procedures developed to identify and quantify individual components in a mixture or of measuring a specific compound or group of positional isomers by means of selected ion monitoring (SIM).

4.5.1. The Mass Spectrometer

The most common technique currently in use for the production of positive ions in the mass spectrometer is electron bombardment that occurs in the ion source. The component of a mixture to be examined is introduced as a gas into the ion source at its operating pressure. The vaporized component is allowed to pass through a slit into the ionization chamber, where it is subjected to a beam of electrons accelerated from a hot filament. The energy of the electron beam can be varied from 0 to over 100 eV.

Since the ionization potential of most organic molecules is in the range of 7–20 eV, the electron beam can supply energy in excess of the ionization potential. The ioniziation potential represents the energy required for removing an electron from the highest occupied molecular orbital.

$$M + e^- \longrightarrow M^{\underline{+}} + 2e^-$$

Since the energy transfer to molecules in the gaseous phase by the bombarding electrons is in excess of the ionizing potential, the excess energy remains in the molecular ion $[M]^{\underline{+}}$ and is used to break the molecular ion bonds so that fragment ions are produced to give a broad range mass spectrum.

The alternative process of capture of an electron by a molecule to form a negative ion radical is less probable, but it does occur.

$$M + e^- \longrightarrow M^-$$

There are various types of mass separation methods. The single- and double-focusing magnetic sector systems are most common in use for standard mass spectrometers. The quadrupole mass spectrometer is the most widely used for GC/MS-coupled systems. Their description and principles are described elsewhere.[26]

4.5.2. Requirements for an Optimal GC/MS Combination

In order to obtain useful information from the mass spectra, it is necessary for the compound introduced into the mass spectrometer ion source to be a single component. This means that the separation power of the GC column should be optimized. If complex mixtures have to be analyzed, high-resolution WCOT column should preferably be used. Stationary phases are chosen, depending on the separation required for the mixture being analyzed. However, when the characteristic masses of two compounds differ from each other, interpretation of the mixed mass spectra is possible. Frequently, excessive GC column bleeding decreases the detection limit for the determination of trace compounds as their low-intensity ions overlap those of the stationary phase. Further, this may result in severe decreases in stability, sensitivity, and resolution.

4.5.3. Requirements for the Mass Spectrometer

The requirements for the mass spectrometer in a GC/MS system are more complicated. They have been reviewed in the literature.[26-28]

4.5.4. Vacuum Technology

The mass spectrometer must be operated under high vacuum so that ion molecule reactions and peak broadening can be avoided. It is mandatory that both the ion source and analyzer pumps have sufficient pumping capacity. Therefore, they are equipped with a differential pumping system designed to accept the total gas chromatographic effluent.[29] Very good pumping capacity is achieved with a turbomolecular pumping system. They are being increasingly used in mass spectrometric systems.

4.5.5. Ion Source Optics

The ion source optics has to produce and maintain a well-defined peak shape and resolution during the entire GC run. This is important in obtaining reliable results in quantitative trace analyses of environmental samples with selected ion monitoring. The mass spectrometer can be considered to be the ultimate in specific detectors since a mass spectrum is completely unique. The ions can be produced in many ways, such as by electron impact (EI) using a beam of electrons in the vapor phase; by chemical ionization (CI) where the compound under study is ionized by reaction with a set of reactant ions; by charge transfer by

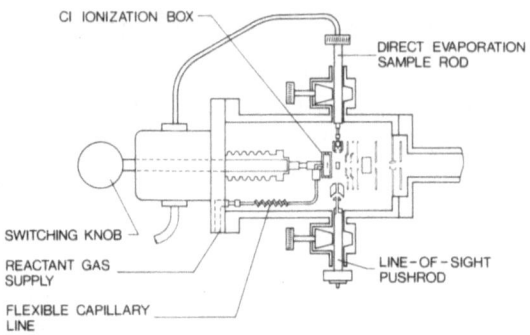

FIGURE 4.5. Combined EI–CI ion source. (Courtesy of Finnigan MAT Corp.)

transferring H^+ with the consequent formation of even electron ions; and by field ionization (FI) and field desorption (FD) techniques from the emitter directly. As an example, the combined ion source is shown which can be used alternatively for EI (10^{-5} torr pressure) or CI (1 torr pressure) mode of operation. The changeover from the one mode of operation to the other can be performed quicky from outside the housing without breaking the vacuum. Figure 4.5 shows the functions and principle of the operation for changeover of modes on a Finnigan MAT 311A ion source.

Changeover to the CI mode is effected by pushing in the switching knob up to a stop. The more closed CI ionization box fits into the EI ionization chamber and effectively gives the required pressure of sample.

Arrangements and function of the electrodes for EI and CI modes are shown in Fig. 4.5. The electrodes common to both modes of operation are named in the middle of Fig. 4.6.

In the CI mode of operation the molecules of the sample are ionized by collison of the reactant gas ions. The process of ion–molecule interaction requires a relatively high working pressure ($\sim$1 torr) in the ion-

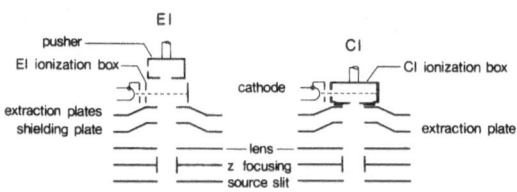

FIGURE 4.6. Typical optical system of an ion source. (Courtesy of Finnigan MAT Corp.)

ization region. In order to fulfil this condition, the ion source is made as gas tight as possible in that mode.

4.5.6. Scanning Time

In order to obtain adequate qualitative information about the purity of a chromatographic peak and to obtain representative mass spectra, the mass spectrometer should be able to register at least six mass spectra per GC peak. A scanning time of 0.1–3.5 sec per decade is required for sharp peaks eluting from WCOT columns.

Fast scanning at high resolution necessitates fast response in the amplifier–recorder system, and the sensitivity is significantly reduced due to the very narrow slit width and number of ions collected. As a rule, 8–12 sec/decade is used at 7000–12,000 resolution.

4.5.7. Repetitive Scan

A repetitive scan is performed by means of the repeated scanning through a preselected mass range by varying the magnetic field strength or the accelerating voltage of magnetic mass spectrometers. For GC/MS work the quadrupole instrument is widely used. A mass scan can be executed in milliseconds because only electric fields and voltages are varied.

4.5.8. Selected Ion Monitoring

In GC/MS the analyst is often interested in obtaining the time intensity distribution of a selected few masses. To meet this requirement, provision must be made available to avoid cycling the system through the complete mass range, but instead focus the mass spectrometer cyclically on a few selected masses of interest. In this way a measurement or integration time is accomplished per mass, which is usually a few magnitudes longer than when scanning. This procedure of selected ion monitoring (SIM) can be realized in the following ways for magnetic mass spectrometers:

(i)   computer-aided adjustment of the mass spectrometer accelerating voltage by means of electrical peak jumping,

(ii) computer-controlled adjustment of the mass spectrometer magnetic field by means of magnetic peak jumping, and

(iii) combined jumping by means of electrical and magnetic peak jumpings (Fig. 4.7).

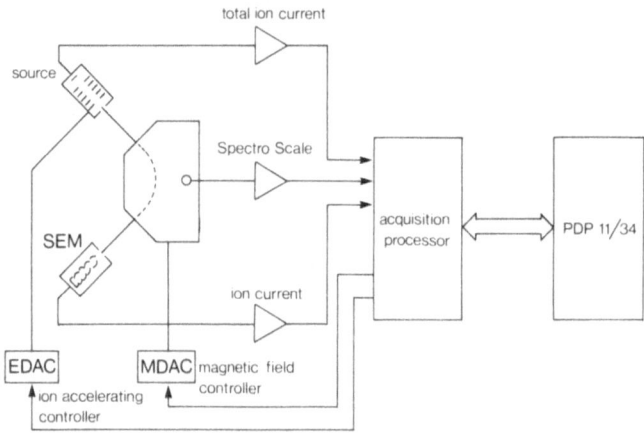

FIGURE 4.7. Computer-controlled selected ion monitoring system. (Courtesy of Finnigan MAT Corp.)

Electrical peak jumping is very fast (jumping time 10–20 msec irrespective of mass) as compared to magnetic jumping (time of 300–500 mscc), but it has the disadvantage of restricted jump range and loss of sensitivity with increasing jump distance. The computer practically compensates the disadvantages of electrical jump and magnetic peak jumping by using combined peak jumping.

Since an accelerating voltage can be switched between preselected values considerably faster and more reproducibly than a magnetic field, because of hysteresis and magnet inertia in magnetic mass spectrometers, the accelerating voltage is varied preferentially. In the case of quadrupole mass spectrometers, selected ions are detected by varying the quadrupole field, which only involves changing electric fields and voltages.[30-32]

In methodological investigation the addition of a compound labeled with a stable isotope is used to increase the sensitivity of a GC/MS system and simultaneously to perform a quantitative determination. This type of analysis can be performed by single- or multiple-ion detection techniques in the SIM mode of operation. The detection sensitivity for an ion signal rises to the degree in which the proportion of the measuring time of the electron multiplier signal in the total measuring time is increased. Both in the detection of one ionic species and in the use of a multiple detection an optimum sensitivity is achieved.

Labeled compounds containing stable isotopes have become commercially available. Such compounds have proven to be excellent as internal standards for quantitative GC/MS studies. Since only microgram quantities of the standard at most are needed for quantitative determination and quality assurance, their use is quite acceptable.

4.5.9. Computer Compatibility

The time required for recording and processing of the data in GC/MS analysis is considerably reduced by coupling the GC/MS to a computer. This permits the treatment of a large number of samples and the generation of data, which is not possible without the computer. Recent developments permit the data system to monitor and control the analytical parameters. This provides a simpler and more reproducible operation. The application of computers to GC/MS can be useful in both high-resolution and low-resolution modes of operation.[32]

In general, the various procedures and modes of operation utilized in GC/MS applications are shown in Fig. 4.8. The scheme will be more complete when these techniques are connected with the data system. Figure 4.9 shows a block diagram of the GC/MS–computer system.

4.6. OPERATIONAL MODES OF AN ION SOURCE

Mass spectrometric modes of operation have provided the analytical environmental chemist with the tools necessary to solve complex environmental problems that previously defied trace analysis because of the low volatility or the high molecular weight of the material or polar functional groups. A brief discussion of modes of operation principles is given.

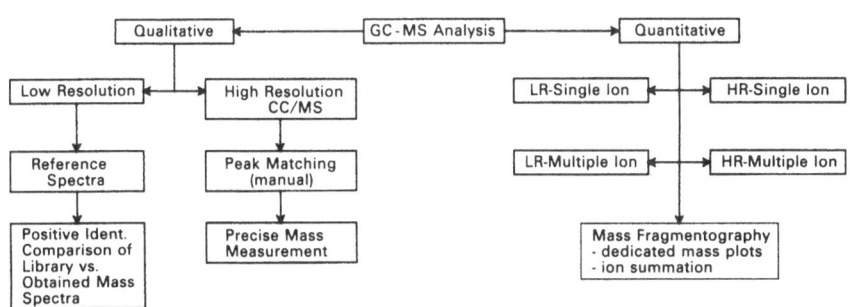

FIGURE 4.8. Summary of GC/MS operational capabilities.

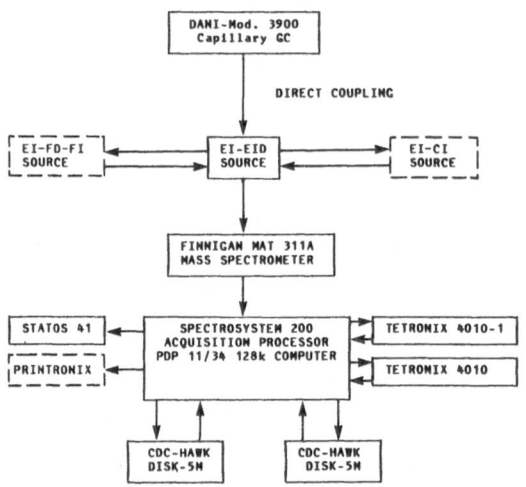

FIGURE 4.9. The high-resolution GC/MS computer setup.

4.6.1. Electron Impact Ionization

The most common method used to effect ionization of organic molecules is based on an exchange of energy during electron impact (bombardment) of neutral gas molecules resulting in a Franck–Condon transition, which produces a molecular ion, an odd-electron species usually in a high state of vibrational and electronic excitation.

Usually, the electron energy may be varied with the EI source, thus permitting one to determine the ionization potentials as well as the appearance potentials of organic molecules. Molecular ionization is initiated when the electron energy is greater than the ionization potential of the molecule. If the electron energy is increased, ionization efficiency also increases. It has been found that maximum ionization efficiency occurs at an electron energy of approximately 70 eV for most organic compounds. Fragmentation and dissociative ionization occur, and the resulting mass spectrum indicates characteristic decomposition pathways of the particular compound under study.

4.6.2. Chemical Ionization

In chemical ionization mass spectrometry (CIMS) ionization of the compound of interest is effected by ion–molecule reactions rather than by electron impact.[33] Much of the earlier work utilized positive ion–

molecule reactions between reagent ions and the sample.[32] In recent years there has been increased interest in negative ion chemical ionization.[34]

Chemical ionization has been carried out in ion sources capable of operating within the pressure range 0.2–2 torr and with the reagent gas present in at least a 1000-fold excess of the sample. In order to obtain maximum sensitivity, external flow of gas out of the source must be avoided or minimized and the pumping speed available in the ion source must be as high as possible. Commercial ion sources can be redesigned to serve as EI sources at low pressure and CI sources at high pressure, thus allowing the sources to be used in either mode. CI represents a low-energy process, so less molecular fragmentation is observed. The fragmentation pattern of a molecule can be varied by using different reactant gases. A potential advantage of CI is that by means of a suitable choice of the reactant gas the ionization can be confined to a few product ions. For quantitative trace analysis involving selected ion monitoring, the sensitivity of CI is further enhanced.

With molecules of high proton affinities, the formation of MH^+ is the major reaction in methane, isobutane, and ammonia CI mass spectra.

4.6.3. Field Ionization

The field ionization (FI) principle consists of the detachment of an electron from organic molecules in a very high electric field of about 10^7 to 10^8 V/cm. In this method of ionization only a small amount of excitation energy is transferred so that FI mass spectra show a lower relative intensity of fragment ions and more abundant molecular ions in comparison with EI mass spectra.[35] If special methods of ion detection are used (i.e., a multichannel analyzer for ion counting), even with relatively weak ion currents, it is possible to perform quantitative trace determinations at low pg/μl levels.

4.6.4. Field Desorption

In the field desorption mode a sample is placed directly onto the emitter surface. Field desorption permits gentle ionization of highly polar compounds. The mass spectra frequently show only the molecular ion formed by the attachment of a cation to the molecule. For this reason the molecular ion group carries the total current in a field desorption process, which is a favorable situation in quantitative analysis. However, this technique cannot be utilized in tandem with WCOT gas chromatograph.[35]

4.6.5. Atmospheric Pressure Ionization

Atmospheric pressure ionization mass spectrometry (API–MS) has been found to be a sensitive analytical technique having a high degree of selectivity due to the thermal nature of the ion molecular collisions and the long ion source residence times.[36] The ion source of the mass spectrometer resembles the standard ECD, so that simultaneous monitoring of the ions and monitoring of the electron capture detection response can be carried out for the reaction chamber. Both positive and negative ions are observed. Negative ions observed under API conditions are almost always anions of acids. They may be formed by reaction with a gas phase basic ion or by electron capture reactions. Most pesticides, herbicides, and fungicides, as well as numerous pollutants that are environmental hazards, form stable anions and show an electron capture response. The results indicate that an analytical system of this kind is useful for the detection and quantitation of many environmentally hazardous compounds.[37,38]

4.6.6. Tandem Mass Spectrometry (MS/MS) and GC/MS/MS

A new technique of mass spectrometry has been developed in which tandem mass analyzers are used for separation and identification. In the MS/MS technique several ionic species are generated from a molecule. Ions of a particular mass are selected for fragmentation and the resulting fragments are mass analyzed.

In principle, two mass filters, quadrupoles 1 and 3, form a tandem mass spectrometer, while the center quadrupole 2 produces dynamic focusing for the parent and fragment ions present in the collisionally induced dissociation (CID) region (Fig. 4.10). This transmits the CID product ions with high efficiency to a third quadrupole filter for mass analysis of the resulting CID–mass spectrum. The tandem quadrupole is particularly promising as a routine analytical instrument for GC/MS because of its high specificity and the ease with which it can be under computer control.

It is obvious that increase in the demand for analyses of higher sensitivity and selectivity can be achieved with GC/MS/MS. The combination of WCOT column GC and MS has helped meet this demand by adding specificity over analyses by either individual technique. MS/MS is a much more specific detector than MS alone, thus adding GC to MS/MS is a way to increase selectivity further. High-resolution MS/MS has demonstrated a unique application to important environmental ultratrace analytical determinations with a minimum of sample cleanup.[38–40]

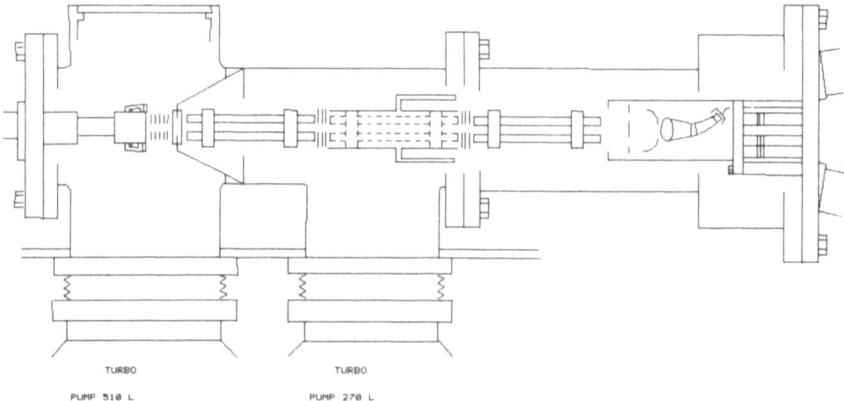

TURBO   TURBO
PUMP 510 L   PUMP 270 L

FIGURE 4.10. MS/MS triple quad system. (Courtesy of Finnigan Corporation.)

4.7. GC/MS INTERFACE

Commercial GC/MS systems are usually interfaced with Pt–Ir capillary tubings as transfer lines between GC and MS. Though generally considered inert, several reports indicate this alloy is responsible for catalytic decomposition of certain substances usually present in environmental samples.[32,41,42] Many different types of interfaces have been reviewed in the literature.[26,43–45]

The ideal properties of a GC/MS interface should be such that:

   (i)   GC separation must not be affected;
   (ii) sensitivity and resolution of the MS system must not be affected;
   (iii) the entire sample should be transferred to the ionization chamber of the mass spectrometer;
   (iv) no chemical changes in the sample should occur during its transfer through the interface;
   (v)   no discrimination against thermally and chemically labile compounds should occur in the interface;
   (vi) no adsorption and memory effects; and
   (vii) efficiency independent of various carrier gas flow rates and temperature.

A narrow-bore WCOT column (0.2–0.35-mm i.d.) can be coupled directly to the mass spectrometer. The total effluent from the WCOT column enters the ion source. The efficiency is 100%. The system is simple, but it is restricted to WCOT columns with flow rates up to 6 ml/

min of helium, depending on the pumping capacity of the pumping system. Vacuum-tight connections between the GC and mass spectrometer are critical. It is not always easy to change the WCOT column.

Schematic presentation of the two most recommended types of interfaces used in GC/MS are shown in Fig. 4.11. The second type of GC/MS interface is represented by the open split coupling.[45] This principle has been significantly improved by using a transfer line of fused silica, which has a very inert surface. It may be used even when the chemical ionization mode is employed without any hazard of electrical discharge from the ion source to the ion source housing or the GC/MS interface. In most magnetic scanning mass spectrometers, the ion source is several thousand volts above ground potential. For this reason methane or isobutane used in CI can explode and damage the ion chamber components or the electronic circuits. Nonconducting materials, such as fused-silica tubing, are superior to a platinum–iridium capillary with regard to electric discharge. A simple, versatile interface is shown in Fig. 4.12.

4.7.1. Open Split

The column fits loosely into the glass capillary tube (B) with the WCOT column end again coming right up to the transfer line capillary (Q). The solvent is elminated by passing helium through line (D) at a pressure of approximately 1 kg/cm^2. After closing valve 11, admission of air is prevented by the continuous supply of helium via line (E).

4.7.2. Direct Coupling

The WCOT column end is brought closely up to the fused-silica transfer line capillary (Q). The solvent is eliminated by vacuum (C) by opening valve 1. The resulting dead volume has to be refilled. This is performed by shutting off valve 1 and opening valve 11 for a short period of time. A helium pressure of 0.2 kg/cm^2 is sufficient (Fig. 4.12).

TYPE OF INTERFACE	FLOW-RATE RANGE mL/min[ll]	EFFICIENCY ENRICHMENT**	OPERATING TEMPERATURE °C	DECOMPOSITION ABSORPTION EFFECT ***
GC OPEN SPLIT — MS a	1–100	1–90% —	< 400%	–
GC DIRECT COUPLING — MS b	≤ 10	100% —	< 400%	–

FIGURE 4.11. GC/MS interfaces.

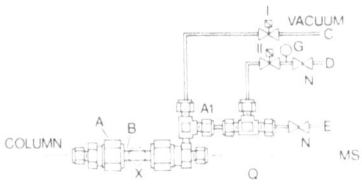

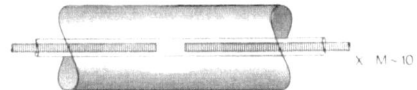

FIGURE 4.12. Direct coupling and open-split between GC and MS. (Courtesy of Hewlett-Packard.)

4.8. PARAMETERS AFFECTING PERFORMANCE OF GC/MS SYSTEM

Manual and computerized operation of GC/MS instrumentation requires daily calibration and resolution checks. Regardless of the superiority of mass spectra collected during an analysis, the final qualitative and quantitative results are limited by the system calibration conditions. An improperly numbered mass fragment is extremely misleading. The normal operating procedure includes calibration once every operating day for the appropriate optimal operational conditions. Routine utilization of a system with or without a data system requires a dedicated operator in order to minimize operational and electronic problems. The sensitivity of a mass spectrometer is very dependent upon the degree of dedicated care the GC/MS system receives.

REFERENCES

1. I. G. McWilliam and R. A. Dewar, *Nature* **181**, 760 (1958).
2. F. W. Karasek and L. R. Field, *Res. Dev.* **28**(3), 42–52 (1977).
3. D. H. Desty, C. J. Geach, and A. Goldup, in *Gas Chromatography-1960* (R. P. W. Scott, ed.), Butterworths, London, 1960, pp. 46–64.
4. L. Onkienhong, in *Gas Chromatography-1960* (R. P. W. Scott, ed.), Butterworths, London, 1960, pp. 7–15.
5. I. G. McWilliam, *J. Chromatogr.* **6**, 110–117 (1961).
6. J. C. Sternberg in *Gas Chromatography 1962* (N. Brenne, ed.), Academic, New York, 1962, p. 231.
7. J. E. Lovelock and S. R. Lipsky, *J. Am. Chem. Soc.* **82**, 431–433 (1960).
8. W. A. Aue and S. Kapila, *J. Chromatogr. Sci.* **11**, 255–260 (1973).
9. E. D. Pellizzari, *J. Chromatogr.* **98**, 323–361 (1974).
10. C. A. Burgett, *Res. Dev.* **25**(11) 28 (1975).

11. H. B. Bente, Hewlett-Packard Techn. Paper No. 75, Avondale, PA (1978).
12. R. J. Maggs, P. L. Joynes, A. J. Davies, and J. E. Lovelock, *Anal. Chem.* **43**, 1966–1972 (1971).
13. P. L. Patterson, *J. Chromatogr.* **134**, 25–37 (1977).
14. W. A. Aue, *J. Assoc. Off. Anal. Chem.* **51**, 682–688 (1971).
15. O. W. Richardson, *Phil. Mag.* **8**, 1–14 (1904).
16. A. Karmen and L. E. Guiffrida, *Nature* **201**, 1204 (1964).
17. K. Abel, *J. Assoc. Off. Anal. Chem.* **49**, 1022–1025 (1966).
18. B. Kolb, M. Auer, and P. Pospisil, *J. Chromatogr.* **134**, 65–71 (1977).
19. M. J. Hartigan, J. E. Purcell, M. Novotny, M. L. McConnell, and M. L. Lee, *J. Chromatogr.* **99**, 339–348 (1974).
20. J. A. Lubkowitz, J. L. Glajch, and B. P. Semonian, *J. Chromatogr.* **133**, 37–47 (1977).
21. C. A. Burgett, Hewlett-Packard Application Note ANGC-2-76, Avondale, PA (1976).
22. W. A. Aue, *Advances in Chemistry*, Ser. No. 104, ACS, Washington, D.C., 1971, p. 39.
23. J. N. Driscoll, *J. Chromatogr.* **134**, 49–55 (1977).
24. J. N. Driscoll, *Amer. Lab.* **9**, 71–74 (1976).
25. M. L. Langhorst, *J. Chromatogr. Sci.* **19**, 98–103 (1981).
26. W. McFadden, *Techniques of Combined Gas Chromatography Mass Spectrometry: Applications in Organic Analysis*, Wiley, New York, 1973.
27. W. D. Lehmann and H. R. Schulten, *Angew. Chem.* **17**, 221–238 (1978).
28. M. C. Ten Noever de Blauw, *J. Chromatogr.* **165**, 207–233 (1979).
29. F. W. Karasek, *Anal. Chem.* **44**(4), 32A–41A (1972).
30. P. H. Dawson, *Quadrupole Mass Spectrometry and Its Applications*, Elsevier, Amsterdam, 1976.
31. W. L. Budde and J. W. Eichelberger, *J. Chromatogr.* **134**, 147–158 (1977).
32. E. O. Oswald, P. W. Albro, and J. D. McKinney, *J. Chromatogr.* **98**, 363–448 (1974).
32. M. S. B. Munson and F. H. Field, *J. Am. Chem. Soc.* **88**, 2621–2630 (1966).
33. A. G. Harrison, *Hydrocarbons and Halogenated Hydrocarbons in the Aquatic Environment* (B. K. Afghan and D. Mackay, eds.), Plenum, New York, 1980.
34. K. R. Jennings, *Mass Spectrom.* **4**, 203 (1977).
35. H. D. Beckey, *Principles of Field Ionization and Field Desorption Mass Spectrometry*, Pergamon, Oxford, 1977.
36. E. C. Horning, D. I. Carroll, I. Dzidic, S. N. Lin, R. N. Stillwell, and J. P. Thenot, *J. Chromatog.* **142**, 481–495 (1977).
37. R. K. Mitchum, G. F. Moller, and W. A. Korfmacher, *Anal. Chem.* **52**, 2278–2282 (1980).
38. T. L. Kruger, J. F. Litton, R. W. Kondrat, and R. G. Cooks, *Anal. Chem.* **48**, 2113–2119 (1976).
39. K. Levsen and H. R. Schulten, *Biomed. Mass Spectrom.* **3**, 137–139 (1976).
40. F. W. McLafferty and F. M. Bockhoff, *Anal. Chem.* **50**, 69–76 (1978).
41. P. P. Schmid, M. D. Müller, and W. Simon, *J. HRC & CC* **2**, 225–228 (1979).
42. W. D. Koller and G. Tressl, *J. HRC & CC* **3**, 359–360 (1980).
43. A. N. Freedman, *Anal. Chim. Acta* **59**, 19–31 (1972).
44. B. J. Gudzinowicz, M. J. Gudzinowicz, and H. F. Martin, *Chem. Instrum.* **8**(4), 225–293 (1977).
45. N. Neuner-Jehle, F. Etzweiler, and G. Zarske, *Chromatographia* **6**(5), 211–216 (1973).

CHAPTER FIVE

SAMPLE PREPARATION

5.1. GENERAL CONSIDERATIONS

Environmental analytical measurements are used for the determination of the transformation and transport of an environmental contaminant and for determination of its environmental concentration levels. A report entitled "Guidelines for Data Acquisition and Data Quality Evaluation in Environmental Chemistry" has been compiled by the ACS Committee on Environmental Improvement and Subcommittee on Environmental Analytical Chemistry.[1] These guidelines represent a thorough and well-designed aid for use by an analytical chemist.

The determination of the quantities and composition of contaminants in the air and water system is a difficult analytical task because of the diversity of the contaminants, the very low levels at which they are found, and the complex matrix of organic compounds from which they must be separated. In general, sampling requires the utmost care and very meticulous attention. Some broad considerations apply to all sampling and should be reviewed before dealing with the specific problems of air and water sampling.

The first prerequisite in any sampling program is to obtain samples that are truly representative of the quantity that is present. The sampling procedure must obtain samples that are identical to the environment at the sampling site prior to sampling. Before deciding on the frequency and type of sampling, it is necessary to decide the accuracy required in the final result.

There would be little benefit in reducing the sampling errors if the analytical errors were large. The errors that occur in sampling are grouped into random and systematic errors. Random errors are the cumulative uncertainties in making measurements, and they are always present to some degree. Systematic errors are much different because they cause all of the results to be shifted away from the true value. In environmental analysis the true values are never known and comparative sampling is difficult, so that systematic errors are particularly troublesome. It is possible to obtain precise, reproducible results that are completely erroneous.

5.2. PLANNING THE ANALYTICAL STUDY

A planning model for the solution to an analytical environmental problem is set on the understanding of the measurement process and a reasonable presumption as to the nature of the problem under study. The model will contain a number of important elements.[2]

5.2.1. Sampling

The quality of analytical data is critically dependent on obtaining representative samples. Analytical results are meaningful if collected samples meet the goals of the monitoring program. A description of sampling sites, techniques, frequency of sampling, and number of samples taken are needed to allow the analytical results to be statistically evaluated. The container for sampling and storage of samples before analysis, including transportation procedures, must be such that alteration of samples does not occur.

Environmental samples are usually heterogeneous, and thus a large number of samples must be analyzed to obtain meaningful data. The number of samples and the quality of the sampling procedure must be well documented for reliability in the results. The number of total samples that need to be run will depend on the kind of information required. If an average compositional value is required, a large number of randomly selected samples are obtained, combined, and homogenized in order to obtain the composite from which subsamples are analyzed. If composition profiles are desired, many samples must be individually measured in replicate analyses.

Environmental trace analysis is performed on samples where the standard deviation of the individual samples is not known in advance and where measurement error cannot be predicted nor can it be assumed to be negligible. In this case the measured values can be used to calculate an overall standard deviation, σ_0, which is related to the standard deviation of measurement, σ_m, and the standard deviation of individual samples, σ_n, by the equation

$$\sigma_0^2 = \sigma_m^2 + \sigma_n^2 \tag{5.1}$$

An approximate value of σ_m can be obtained by a pooling process, using the differences in the measured values of duplicate homogenized samples.[3] Then the standard deviation of the individuals, σ_n, can be calculated. At least seven parallel measurements must be performed in order to obtain a reliable value for the standard deviation. A statistical approach to sampling is possible when the standard deviation of the individual samples is known in advance.[4] The equation

$$N = (Z \cdot \sigma_n/E)^2 \qquad (5.2)$$

is used, where N represents number of samples, Z is a constant, σ_n the standard deviation of individual samples, and E the tolerable error as a mean estimate for the characteristic measured. Assuming that the samples under analysis are expected to have a mean concentration of approximately 100 ppb with a standard deviation of 50 ppb and that the tolerable error in the stated value of the mean at the 95% confidence level ($Z = 1.96$) is not to exceed 20 ppb (20%), the number of samples required is

$$N = [(1.96 \times 50)/20]^2 \simeq 24 \text{ samples}$$

However, when the data needed to calculate the minimum number of samples (N) are not available at the time of sampling, it is often replaced with an empirical approach such as a 7–7–7 rule currently used by the U.S. Environmental Protection Agency. This means that seven field samples, seven field blanks, and seven spiked blanks are to be analyzed along with the controls and calibrating standards. Field blanks of environmental samples are often not readily available. In this case a synthetic field blank is the only alternative and it is prepared in the laboratory.

After a sample has been received, it is homogenized prior to subsampling. The introduction of potentially interfering substances and changes in the contaminant concentration must be avoided. Sample pretreatment may involve sieving, blending, crushing and drying, dissolution, and the addition of preservatives and internal standards. These pretreatments are documented in sufficient detail to provide a complete record of the sample history.

In general, glass, Teflon, and aluminum foil have been proven to be the most suitable materials to come in direct contact with the sample. Plastic containers must be rigidly avoided for samples that will be examined by gas chromatography. Minute traces of the plasticiers are disastrous when using an electron capture GC detector. Aluminum foil or Teflon is generally used as linear material for a bottle or jar cap when the material in the normal cap may contribute impurities.

Water samples may be conveniently taken in glass bottles in which organic solvents such as acetone and hexane are supplied. The molded screw cap has a Teflon liner. If not, an aluminum foil liner may be used.

Environmental samples, such as sediment samples of up to 1 kg may be taken in Mason jars of various sizes. Aluminum foil liners are recommended, and care must be taken to prevent the sample material from contacting the paper liner of the usual metal screw caps.

For fish or wildlife tissues, 1-oz wide-mouth glass bottles (5.16 × 3.16 cm) should be used. These containers are suitable for any sample not exceeding 25 g.

For gaseous samples of ambient air or industrial stack emissions the toxic compounds can be present in either the gaseous or particulate phases. Volatile components are concentrated by passing the gas through solvent, carbon, or porous polymer resins. These compounds are then quantitatively dissolved and introduced into the instrument. After removal of particulates by filtration or cyclonic action, the nonvolatile toxic components are measured by GC/MS or WCOT gas chromatography with and without specific detectors.[5]

Standardization is defined as the response function

$$S = f(c) \tag{5.3}$$

where S represents the measured response signal, which is a function of the given concentration (c) of a component to be analyzed. Where possible, regression analysis consistent with a linear treatment of the function should be carried out.[6] At least five different concentrations of the calibration standard should be measured in triplicate. The concentration range should bracket the concentration of the contamination in the field sample. It is obvious that the standardization must be performed under the same conditions as will be employed during the measurement itself.

5.2.2. Extraction Procedures

Many environmental samples cannot be analyzed directly for residues because the level of the desired contaminant may be too low and the levels of interfering compounds too high. It is usually necessary to extract the desired contaminants from the matrix and to purify and concentrate them prior to determination.

A solvent or mixture of organic solvents may be used for extraction, provided the solvent is at least 80% efficient, selective enough to require a minimum of cleanup, and does not interfere with the final determination. Optimum extraction conditions are found by recovery studies for each analysis. Hexane and hexane–acetone mixtures are typical solvents for nonpolar, fat-soluble contaminants such as organochlorine pesticides. Benzene, dichloromethane, and chloroform are suitable for more polar compounds such as carbamates and organophosphates. Acetonitrile is an excellent general solvent used for partitioning of unknown residues of a wide polarity range.

The more polar solvents remove greater amount of interfering compounds and may complicate subsequent cleanup procedures. Sodium sulfate is almost always added to help extract the more water-soluble compounds. Soxhlet extraction procedure with an appropriate solvent or mixture of organic solvents is the most efficient general method for removal of many environmental contaminants from different matrices. Extraction efficiency can be checked most accurately if the laboratory is able to incorporate radioactive or stable-isotope-labeled contaminants in the sample substrate.

On-site extraction of organic pollutants from a water system has several advantages over collection of samples in the field, which are then shipped to the laboratory for analysis. The problems of transporting large sample volumes are related to the possibility of chemical alteration and adsorption losses on the container surface. In addition, continuous sampling of a water system should be expected to provide a more representative sample. Extraction using polymeric resins has been widely tested by a number of workers both in the laboratory and in the field, yielding successful results for the analysis of organic pollutants in water systems.[7-9]

5.3. CLEANUP PROCEDURES

Cleanup procedures are chosen in terms of practicality, time, reagent and instrumentation availability, and cost involved. The methods chosen must be tested to be sure they allow detection and determination of contaminants at the required sensitivity level, with recovery at least 80% or better and with removal of background interferences. The cleanup required prior to the final determination depends upon the selection of extraction procedure and the analytical method by which the analyte will be determined.

Injection of uncleaned samples into a gas chromatographic column can cause spurious peaks, damage to the peak resolution and efficiency of the WCOT column, and loss of detector sensitivity. Extracts containing fatty material can greatly shorten the lifetime of a WCOT column. Depending on the extent and composition of the matrix of the coextractive materials, partition between immiscible solvents, gel permeation or liquid chromatography, distillation, selective chemical reactions, or reaction gas chromatography are most often used for cleanup of various types of samples.

5.3.1. Solvents and Reagent Purity

The solvents used in environmental trace analysis by WCOT column gas chromatography must be of very high purity. The purchased solvents and reagents should bear the manufacturer's designation of "pesticide quality, distilled in glass." Even these solvents must be checked for assurance of quality. In ultratrace analysis at ppt level, distillation through an all-glass still is practically mandatory. The electron capture detector especially requires solvents that are free of impurities, giving no detector response at the electrometer attenuation normally employed.

A test for substances causing interference is performed in the following manner. In a precleaned 500-ml Kuderna–Danish concentrator equipped with a Snyder column and a 10-ml evaporative concentrator tube, 300 ml of the solvent is evaporated in a hot water bath to 5 ml. Usually, 2 μl of this concentrate is injected into the WCOT column, and the resulting chromatogram is evaluated for compounds eluting in the retention range of the analysis. If no peaks elute in the retention range of the compounds of interest, the purity of the solvent is adequate. If a peak elutes at the retention time of one or more components of interest, the solvent is not acceptable. Solvents used for partition cleanup of hexane extracts include acetonitrile, dimethyl formamide, and dimethylsulfoxide.

Acetonitrile should be checked for impurities employing a simple test in which vapors from acetonitrile turn moistened litmus paper blue if the reagent is contaminated. Ethyl ether must be free of peroxides. To do this, 10 ml of ethyl ether is first tested with 1 ml of fresh 10% potassium iodide solution. The mixture is shaken and let stand for 10 min. No yellow color should be observed in either layer.

5.3.2. Quality of Adsorbents

5.3.2a. Florisil. The most common adsorbent used in cleanup procedures for environmental samples is Florisil. The pesticide grade known as "PR grade" is available from a number of distributors. It is used for pesticide residue analysis. Its quality is checked at the producers quality control laboratory for activity to ensure uniformity. However, it is recommended that the elution and recovery characteristics for the contaminants of interest be determined by the analyst. Florisil should be stored in an oven at 130 °C at least 16 hr before use.

Florisil column chromatography is used for the cleanup of chlorinated hydrocarbons, organophosphates as well as triazines, carbonates, and others from various matrices, by elution with solvents of increasing polarity. The solvent system consists of n-hexane (or petroleum ether), and its mixture with 6, 15, and 50% of ethyl ether in n-hexane, as well as 100% ethyl ether. These elution mixtures cover most pesticides, including the phenoxyalkanoic acids and their esters.[10]

Before using a Florisil separation, it is important to standardize the Florisil in order to get the correct elution order of the pesticides or PCBs. The Florisil can be standardized for adsorptivity by adsorption of lauric acid, where the excess nonadsorbed acid is determined by titration.[11] The lauric acid value derived determines the amount of Florisil that gives a certain activity and capacity. If properly activated, the Florisil column shold elute common pesticides in specific fractions for GC–ECD determination (see Table 5.1). More complete details in regards to elution order of many compounds are contained in specialized analytical manuals.[10,12]

Florisil cleanup is performed in a glass column 22-mm i.d. $\times$ 20–30 cm long with a solvent reservoir at the top. The outlet should have a coarse glass frit and a stopcock to regulate flow. The column is packed with a slurry of Florisil in petroleum ether to about 10 cm of column length when settled, or with an exact amount as calculated from the lauric acid value.[11] About 1 cm of anhydrous sodium sulfate is placed over the Florisil to take up traces of water that may be left over from the sample.

Kuderna–Danish assemblies of 250 ml are placed under each Florisil column with 10-ml graduated tubes. A 200-ml portion of elution solvent of petroleum ether-ethyl ether (94 + 6 v/v) is employed at a rate of 5 ml/min. The eluate is collected in two separate Kuderna–Danish assemblies. At the instant the solvent level reaches the top of Na_2SO_4 layer,

TABLE 5.1
Compounds Contained in Fractions Eluted from a Florisil Column

6% ethyl ether in petroleum ether				
Aldrin	Hexachlorobenzene	DDE	PCB	TCDD
Heptachlor	o,p'-DDT	Mirex	Disyston	
Lindane	p,p'-DDD	Heptachlor epoxide	Trithion	
Methoxychlor	p,p'-DDT	BHC	Ronnel	

15% ethyl ether in petroleum ether		
Dieldrin	Ethyl parathion	Methyl parathion
Endrin	Diazinon	
Parathion	Thiodan I	

50% ethyl ether in petroleum ether	
Thiodan II	Malathion

another portion of petroleum ether–diethyl ether (85 + 15 v/v) is employed. The eluate is collected in another two Kuderna–Danish assemblies. The elution is continued with 200 ml of petroleum ether-diethyl ether (50 + 50 v/v).

Eluents are concentrated on a steam bath to approximately 2–5 ml. The remaining solvent is evaporated under a nitrogen stream to 1 ml. The environmental samples are diluted with isooctane to exactly 5 ml. The sample is then ready for GC analysis.

5.3.2b. Silica Gel. Silica gel is a widely used chromatographic adsorbent used for cleanup. It is commercially available in a wide range of forms. The literature contains numerous applications of its use in adsorption chromatography. Its unsurpassed capacity for both linear and nonlinear isotherm separations and almost complete inertness toward labile components makes it very useful.

Silica gel deactivated with water has been employed for cleanup and fractionation in various environmental determinations. Silica gel is first activated at 175 °C for 48 hr. Afterwards, 20 ml of water is added to 100 g of adsorbent stored in a tightly capped Teflon-lined screw top jar and the contents is mixed on a rotary mixer for 2 hr. Silica gel prepared in this manner can be stored in a capped jar for at least 2 weeks with no change in adsorptivity.

5.3.2c. Alumina. Alumina is a good all-purpose adsorbent and is commercially available in neutral, basic, and acidic forms. Basic alumina has a reputation for altering acids, which react to form salts on its basic surface. Esters and anhydrides also undergo saponification. Acetone also reacts in the presence of basic alumina. Isopropyl ether as solvent on alumina has also been reported to form high-boiling products. This fact should be kept in mind when minute quantities of environmental pollutants are studied.

5.3.2d. Activated Carbon. Two different types of activated carbon are used: graphitized (nonpolar) carbon, which is prepared by high-temperature activation, and oxidized (polar) carbon, which results from low-temperature oxidation.[13] The first type exhibits little selectivity for different types of compounds, and adsorption is determined largely by the size of the sample molecules. The second type selectively adsorb polar molecules in preference to larger nonpolar molecules.

5.3.2e. Celite. Celite 545 represents a diatomaceous earth, low surface area adsorbent that has been occasionally employed for the separation of polar molecules. This adsorbent has been recommended for labile samples and is frequently used in mixture with other adsorbents, either to improve solvent flow rate or decrease adsorbent activity. It must be pretreated by washing with 6 *M* HCl while heating on a steam bath, rinsing with water until neutral, and washing with several solvents.

5.3.3. Group Selective Adsorbents

5.3.3a. Ion Exchange Resins. The ion exchange resins have received a great deal of attention as adsorbents of organic contaminants in environmental analysis. The XAD 1 to 5 resins are based on polystyrene-divinylbenzene copolymers and are nonpolar, whereas XAD-7 and XAD-8 are polar polymethacrylate resins.[7,14]

Prior to their use in the collection and concentration of organic compounds from water and air systems, the resins must be cleaned to remove impurities or nonpolymerized oligomers that will be extracted when the pollutants are desorbed from the resin. The extraction step involves a 6-hr Soxhlet extraction with methanol followed by a 6-hr Soxhlet extraction with diethyl ether. Afterward, the resin is stored under methanol in a glass bottle until it is to be used. Storage of the resin under methanol prevents adsorption of organic pollutants from the atmosphere and also prevents the resin from drying out. A resin blank must be run in parallel with the sample. An apparatus for extracting organic contaminants from water is shown in Fig. 5.1. A silanized glass wool plug (B) is inserted near the stopcock. The methanol slurry of XAD resin (D) containing 2 g of dry resin is added. A second plug of pre-extracted silanized glass wool is placed over the resin, the methanol is drained, and the resin is washed with 3 × 20 ml portions of water. Care should be taken to drain the solvent only to the top of the resin bed. The reservoir

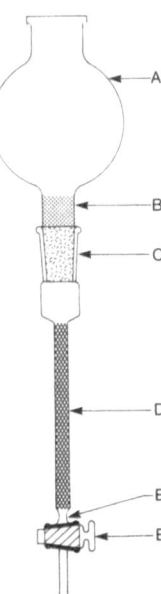

FIGURE 5.1. Apparatus for extracting organic pollutants from water. (A) Reservoir; (B) silanized glass wool plugs; (C) 24/40 joint; (D) 0.6-cm-i.d. × 10-cm-long glass column containing 40–60 mesh XAD-2 resin; (E) PTFE stopcock.

flask has a capacity up to 5 l. The water sample is introduced into the reservoir and the top of the reservoir is connected to a tank containing ultrapure nitrogen. The flow rate through the resin bed is adjusted to 30–50 ml/min by adjusting nitrogen pressure to approximately 0.1 kg/cm^2. After the water has drained from the column, the sample is extracted with diethyl ether. First, the reservoir is washed with two 10-ml portions of diethyl ether and the contents are retained on the column for 10 min. After this, the ether is drained into a test tube and an addditional 5 ml of diethyl ether is passed through the column. The column bed can be regenerated by passing 30 ml of methanol through the column and shaking the column to remove entrapped air. The stopcock is then closed, 15 ml of methanol is added, and the column is ready for use again.

The microdistillation method using the Snyder distillation assembly is superior to the nitrogen gas stream method.[7] The volume of the sample collected in Kuderna–Danish evaporator is reduced to a predetermined volume (e.g., 1 ml). A small amount of isooctane (0.5 ml) is added, and the remaining solvent is evaporated to 0.5 ml. The final volume is adjusted with isooctane to 1 ml. The scale drawing of the concentration apparatus is shown in Fig. 5.2. A miniversion of the procedure using a minicolumn has been described.[15]

5.3.3b. Tenax. Tenax is a porous polymer based on 2,6-diphenyl-*p*-phenylene oxide. The physical characteristics of Tenax are documented in the literature.[16] Its applications for the extraction of organic micropollutants in water cover pesticides, polycyclic aromatic hydrocarbons,[17] PCBs,[18] and atmospheric pollutants.[19,20]

The extraction of pesticides from surface waters by adsorption on Tenax yields results equivalent to those obtained by the liquid–liquid extraction procedure when applied to drinking and surface waters which completely or almost completely lack solid matter. In instances where

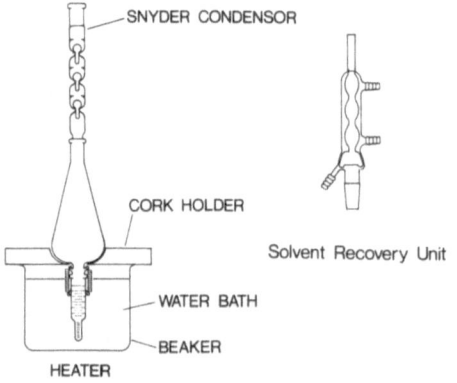

SNYDER CONDENSOR

CORK HOLDER

Solvent Recovery Unit

WATER BATH

BEAKER

HEATER

FIGURE 5.2. Kuderna–Danish evaporative concentrator with the Snyder column.

TABLE 5.2
Separation Scheme of PAH from Open-Pore Polyurethane Foam Extract

Steps	Description
1	Concentrate foam extract down to 10 ml of cyclohexane.
2	Extract is washed with 2 × 60 ml portions of 4 : 1 methanol : water.
3	Cyclohexane layer is retained; methanol + water layer is discarded.
4	Retained layer is washed with 2 × 60 ml portions of distilled H_2O.
5	Cyclohexane layer is kept.
6	Cyclohexane layer is extracted with 3 × 20 ml of DMSO. DMSO combined layer is retained; cyclohexane layer is discarded.
7	120 ml of dist. H_2O is added, then the contents are extracted with 2 × 40 ml portions of cyclohexane. Cyclohexane layer is retained; H_2O–DMSO layer is discarded.
8	The cyclohexane layer is dried by passage through anhydrous sodium sulfate.
9	The total volume is concentrated to 5 ml.
10	Florisil cleanup column is used and the sample is eluted with 125 ml of benzene.
11	The eluate is concentrated to 500 μl.

suspended solids are present, the water samples should be filtered and both the filtered water and the residue on the filter analyzed.

5.3.4. Open-Pore Polyurethane

Polyurethane foams have been used by a number of investigators to recover PCBs, pesticides,[21,22] and phthalate esters from water.[23] In site preparation of the open-pore polyurethane columns containing an OH–NCO ratio of 2.2 have a higher capacity for polycyclic aromatic hydrocarbons (PAH) than lower OH–NCO ratios.[24] The recovery factor for benzo[a]pyrene between 2 ppt and 25 ppb is between 84 and 87%. The separation scheme for PAH from polyurethane foam is shown in Table 5.2.

5.4. LIQUID–LIQUID EXTRACTION

Liquid–liquid extraction of organic pollutants from water has been a very effective method for removing contaminants in addition to thermal desorption and headspace extraction methods.[25] Methylene chloride as the extraction medium is used predominantly since it is available as a highly purified low-boiling point solvent, relatively inert, and readily separated from the water phase.

When quantitative analytical results are needed, a correction factor must be applied to compensate for incomplete extractions. This is accomplished with two sequential extractions on the same water sample.

The second addition of methylene chloride can be reduced to 10 ml since the water is already saturated with methylene chloride. If the integrated peak area for a component in the first and second extractions are represented by A_1 and A_2, and the extraction efficiency is assumed to be the same for the first and second extraction, the total area (A) in the water sample for a component can be calculated from the equation:

$$A = \frac{A_1^2}{A_1 - A_2} \qquad (5.4)$$

Satisfactory results are obtained at the mg/l level, even when the percentage recovered in the first extraction is less than 30%.

When a water sample is extracted several times with a solvent, the extracts are combined and analyzed as a composite. Perhaps, the best approach would be to prepare and analyze standard mixtures as external standards with an unknown sample. However, this would be very time consuming for multicomponent samples. The potential of liquid extraction for the recovery or organic pollutants from water at concentrations of 5 ng/l has been investigated.[26] In this method 1 l of water is extracted with 200 μl of pentane to yield a pentane extraction of approximately 200 μl, which is then made up to 250 μl. To achieve a reasonable extraction with so small a volume of solvent, a solvent of very low aqueous solubility is necessary and the water sample is presaturated with the solvent.

An extraction flask that consists of a two-neck 1 l flask with a capillary tube sealed to a shortened center neck and employs 200 μl of hexane was recently described.[27] The extraction requires only 2 min per extraction. The micromethod repeated three times provides 90% recoveries for pesticides, hydrocarbons, and phthalates. A preconcentration factor larger than 10,000 can be achieved.

There are various types of continuous-liquid extraction apparatus available. An integrated sampler, which is in principle a mixer–settler and combines two extractors in series, has been described for the extraction and analysis of PCB in estuarine water in a dredging sludge disposal area.[28]

5.5. LIQUID CHROMATOGRAPHY

The principle of liquid chromatography has been employed to concentrate and analyze organic pollutants from water systems. In this technique organic pollutants are enriched on an adsorption column, directly

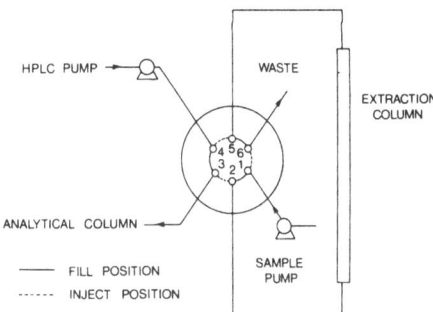

FIGURE 5.3. Connection of the extraction column to the loop injection valve fill position (solid line); inject position (dotted line).

eluted from this column onto an analytical column, and separated by reversed-phase high-performance liquid chromatography (HPLC).

The advantages of this technique are that the volume of sample is minimized since all organic pollutants may be extracted theoretically on the precolumn containing a selective adsorbent. Afterward, they are transferred to the analytical column, where they are separated, partially identified, and collected for further analyses. Sample handling connected with extraction and concentration is eliminated.[29] The extraction and analytical apparatus is shown in Fig. 5.3.

An automatically controlled six-port valve first directs the water sample through a precolumn packed with an adsorbent and then out to waste. After the desired volume of water is sampled, the six-port value is automatically switched so that the eluant back flushes the pollutants adsorbed on the precolumn onto the analytical column.

A displacement chromatography method for enrichment of phenols and PAH from water samples can be used.[30] The water sample is pumped through the enrichment column containing LiChrosorb SI-60, where the trace organics are adsorbed. The preconcentrated compounds are eluted to the end of the column by a displacer solvent, which is either a pure solvent or a solvent component of high concentration. The limiting factors of the enrichment process are elution volume, concentration, and the loading capacity of the column.

5.6. HEADSPACE ANALYSIS

Headspace chromatographic techniques have been widely used.[31-33] In headspace analysis an inert gas is bubbled through or over the water sample. The inert gas, water vapor, and components contained in the water sample are stripped. They pass through a trap that may be an open cyogenic one or one containing a solvent. They may also be directly concentrated on a chromatographic column. The effective degree of enrichment, E_{ef}, can be defined as

$$E_{ef} = qC_{ic}/C_{io} \qquad (5.5)$$

where C_{ic} and C_{io} are the concentrations of component i in the concentrate (c) and in the original material (o), and q is the volume fraction of the total concentrate that is introduced onto the GC column.

Distribution of the contaminant in the gas-condensed phase can be studied by means of techniques in which a sample of the gas taken from over the condensed phase in a closed static equilibrated system is analyzed, or by techniques where a stream of gas is passed through the condensed material and the gaseous effluent is analyzed for the components stripped from the matrix. The concentration of the component in the former system does not change with time after the system has reached equilibrium. In the second case the component concentration descreases continuously with time in both phases, approaching zero asymptotically. It can be assumed that even at these nonstationary conditions the gaseous phase in the bubbles passing through the condensed phase is practically equilibrated with the matrix.

Direct analysis of headspace gas samples has been employed for the determination of halogenated hydrocarbons in water.[34] The system is equilibrated under vacuum at an elevated temperature. The pressure in the system is equalized with the ambient pressure, and a 5-μl sample of the headspace gas then is subjected to chromatographic analysis. Chloroform and other halogenated compounds in water can be determined at a concentration range of 0.1–10 μg/l.

In experiments with headspace gas samples, it was observed that if the septum of the GC injector is not leakproof, the reliability of results may be impaired due to losses of samples or adsorption. An improvement for conducting headspace gas analyses without using a septum is seen in Fig. 5.4. The sample is drawn into a calibrated sampling loop using vacuum. By turning the valve to a second operating position, the contents are introduced into the gas chromatographic column.[35]

A method of headspace analysis based on multiple extraction of a fixed sample of liquid with a gas has been reported.[36] The method was applied to the determination of hydrocarbons in water. In this method a 25-ml sample of water is equilibrated with 25 ml of helium in a 50-ml injection syringe at an elevated temperature. The entire volume of the gas sample is introduced from the syringe into the sampling loop of the gas chromatograph. In order to determine the contents of hydrocarbons

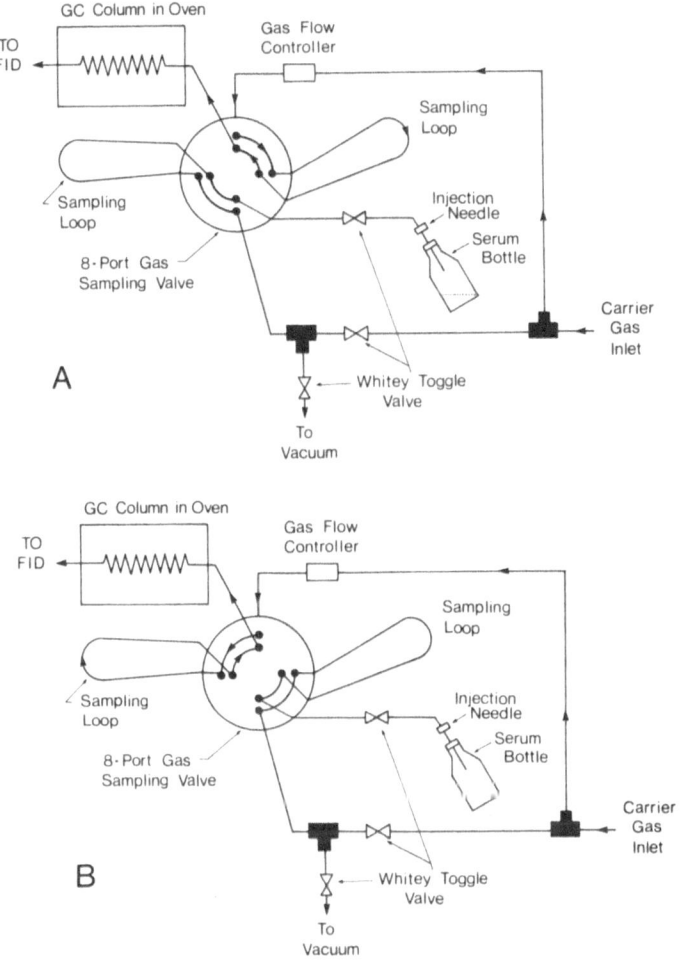

FIGURE 5.4. Gas-sampling device for headspace gas analysis.

in the gas, another 25-ml volume of pure helium is drawn into the syringe
and the whole procedure is repeated several times. The initial concentration of hydrocarbons in the original sample of water is calculated from
the dependence of the concentration of hydrocarbons in the headspace
gas samples on the serial number of the extraction step. It is claimed
that alkanes and cycloalkanes can be determined at concentration as low
as 3 ppt in water. Headspace gas analysis involving quantitation by a
standard addition method and WCOT column gas chromatography with

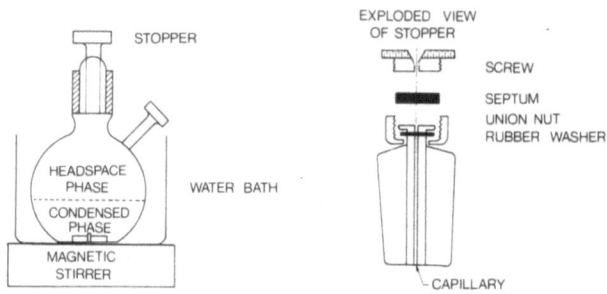

FIGURE 5.5. Equilibrating flask for headspace gas analysis.

splitless sampling permits the reliable determination and identification of ultratrace amounts of volatile hydrocarbons in water and air.[37,38] By employing conventional syringes for the headspace gas sampling and with manual measurement of peak heights, an average error below 10% of the value being determined appears possible for concentrations in the condensed phase down to ppb level. The equilibration devices used are shown in Fig. 5.5. An equilibration time of 30 min is sufficient at 40 °C. The reproducibility corresponding to a relative standard deviation of 10% is very good. The results obtained with these model systems can be generalized for most problems concerned with trace amounts of volatile hydrocarbons.

5.6.1. Closed-Loop Stripping Techniques

Trace contaminants of headspace gas can be concentrated by trapping them in a short column packed with a sorbent. During the stripping process concentration of the components in the gas leaving the system is continuously decreasing.

Measurement of volatile and semivolatile, intermediate molecular weight organic pollutants were analyzed in water at ng/l (ppt) level.[39] The stripping apparatus employing a closed-loop inert gas stripping is seen in Fig. 5.6. The sample flask (1–4-l volume) is stoppered with a ground-glass joint. The headspace gas is recirculated by means of a pump and pases through the sintered glass and a trap containing an activated carbon trap. As the gas passes through the sample, organic components are purged from the water into the headspace gas. The trap then adsorbs these components from the gas. The water bath temperature is controlled at 30 °C, and the preheater temperature is maintained at 80 °C to prevent condensation of vapors in the carbon trap. The trap holder is maintained

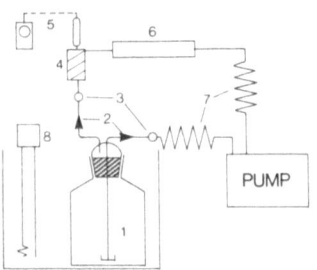

FIGURE 5.6. Closed circuit for stripping with inert gas. (1) Coarse glass frit; (2) PTFE bowl joints; (3) fused glass metal connectors; (4) aluminum heating mantle; (5) heater (15 W) and time relay; (6) filter holder; (7) tubing (stainless steel); (8) temperature control of water bath.

at 40 °C and contains 1.5–5 mg of activated charcoal. At 50 ng/l 1-chloroalkanes are used as internal standards (except for 1-chlorooctadecane which requires 250 ng/l due to its lower recovery). The concentrate is recovered by extracting the trap with 5–30 μl of carbon disulfide or methylene chloride.

The reliability of the quantitative results depends on the efficiency of stripping the components from the water and the adsorbent as well as the losses caused by adsorption or condensation. This technique yields high concentration enrichment efficiency while covering a wide range of compounds. The danger of introducing artifacts is minimal, and the possibility of bringing the system into equilibrium makes it feasible to apply methods of equilibration trapping.

5.6.2. Preconcentration of Contaminants Using Automatic Purge and Trap Devices

The Hewlett-Packard HP7675A Purge and Trap system represents one of many systems available. The system employs the automatic purge–trap–desorption sequence. The quantitative precision is better than 5%. It requires a sample size of 1–50 ml. Preconcentration is achieved on Tenax or other similar adsorbents. Automatic control of the entire analytical sequence is done, including mass spectral library search where results from the GC/MS run are interpreted qualitatively.[40]

An application of this system, which uses special electrolytic stripping cell for the determination of hydrocarbons in sea water by means of the hydrogen generated stripping of a known volume of water (1–3 l), has been described.[41] In the main compartment of the cell, several liters of hydrogen bubbles are passed through the water sample. The hydrogen is generated on a electrode with a large surface area. The amount of hydrogen evolved can be accurately controlled by the total amperage of time according to the following equation:

$$g = \frac{M/z}{96,487} \cdot I \cdot \tau \tag{5.6}$$

where M is the weight of one grammole, Z is valence, 96,487 is the Faraday constant, I is the current, and τ is time.

The most volatile components are almost quantitatively stripped out of the sample by the first hydrogen. If the concentration of pollutants are at the ppb and ppt levels, the components stripped out of the liquid should be trapped in a closed-circuit system containing a few milligrams of active carbon. No artifacts are introduced into the system by the stripping gas. The rate of stripping dW/dt expressed for the total mass W_i of component i in the system can be described as

$$\frac{dW_i}{dt} = -W_i \cdot \frac{F}{V_g + K_i V_L} \tag{5.7}$$

Where F is the volume flow rate of the stripping gas, W_i is the instantaneous total mass of component i in the system, V_g and V_L are the volumes of the gaseous and condensed phases, and K_i is the distribution constant. Solving the above equation for $t = 0$, and W_i^0, gives

$$\frac{W_i}{W_i^0} = \exp\left(-\frac{Ft}{V_g + K_i V_l}\right) \tag{5.8}$$

Where W_i^0 is the initial total mass of component i in the system.

The half-time of the stripping process, $t_{1/2}$, is calculated from

$$t_{1/2} = \ln 2 \cdot \frac{V_G + K_i \cdot V_L}{F} \tag{5.9}$$

The time necessary to strip 95% of component i out of the system is

$$t_{0.05} = 3(V_g + K_i V_l)/F \tag{5.10}$$

These equations assume that the concentration of component i in the stripping gas leaving the condensed phase is in equilibrium with the concentration of this component in the condensed phase. In water most of the components probably will remain in the liquid. Under these circumstances, quantitative analysis is feasible only by calibration using a reference system with the same matrix as in the system analyzed at constant-stripping flow rate, time, and temperature.

A different type of vapor-stripping apparatus is used for volatile organic analyses,[41] and it operates as follows. A helium-purging gas is precleaned by passing through molecular sieves at liquid nitrogen temperature. The sample compartment holds 3.5 l of water. The sample is held at 80 °C by a hot water bath circulator, which allows a wide range of volatile components to be analyzed. The provision of a condensor above the sample holder helps to condense any water vapor purged. Duplicate Tenax adsorbent traps are provided for trapping volatile components from water. The sample is purged at 80 °C with 60 ml/min purified helium for a period ranging from 20 to 60 min. The Tenax traps are then manually transferred into the injection port of a gas chromatograph at 250 °C and analyzed both by GC–FID and GC–MS techniques.

5.7. THERMAL DESORPTION METHODS

An advantage of thermal desorption methods is that all of the components from water or air samples that have been adsorbed on a trap column are transferred onto a WCOT column. This is in contrast to an extraction procedure in which the volume of the extracted sample is reduced to a certain known volume and an aliquot of about 1:1000 is used for analysis. Thus, by using thermal desorption, a gain of a factor of 1000 in sensitivity may be achieved. Another advantage of thermal desorption methods is that volatile components are not lost in the extraction and preconcentration steps, and so one has the choice of either utilizing the gain in sensitivity or reducing the sample volume. A major difficulty in the analysis of water samples by thermal desorption methods, which is not so significant in the analysis of air samples, is that certain amounts of water remain in the interstitial volume of the collection trap packing, and this is transferred onto the WCOT column with the sample. This problem may be solved by a precolumn containing calcium carbide or aluminum hydride. The water would then react with the precolumn packing to produce acetylene or hydrogen. These gases would not interfere with the chromatographic process.

5.8. STEAM DISTILLATION

The steam distillation method is a well-established technique for separating components on the principle of differences of their vapor pressures over water. Conventional steam distillation techniques use an external stream generator that produces and delivers steam into the

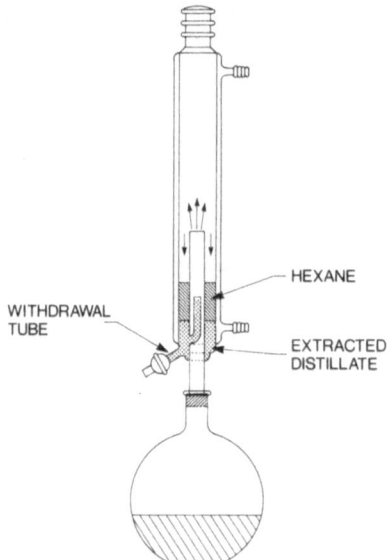

WITHDRAWAL
TUBE

HEXANE

EXTRACTED
DISTILLATE

FIGURE 5.7. Nielsen–Kryger steam distillation apparatus and solvent extraction apparatus.

sample and collects the volatile components by condensing the steam
and components in a cooled flask. Although this method is commonly
employed in flavor and drug analysis,[42,43] the large empty volume of the
glassware and the low collection efficiency have prevented the use of
steam distillation as a quantitative technique in the analysis of trace
concentrations of less volatile organic contaminants. The vapor pressures
of many trace chemicals are greater than those of the high molecular
weight lipids in fish, sediment, and biota. Because exhaustive solvent
extraction techniques remove lipids, waxes, and pigments as well as the
trace contaminants, extensive cleanup separations are necessary before
the extracts can be analyzed.

A modified Nielsen–Kryger steam distillation technique has been
developed for the extraction of pesticides and industrial chemicals from
water, sediments, and tissue.[44] The apparatus provides for the simultaneous distillation and extraction of trace contaminants from water by a
small volume of organic solvent. The exhaustive steam distillation and
solvent extraction apparatus is shown in Fig. 5.7. The extract is generally
suitable for direct GC analysis without the time-consuming cleanup and
concentration steps. The authors have employed this technique for quantitative analysis of chlorinated benzenes in sediments at a 1-ppb level.[45]
Recoveries have ranged between 75 and 99% for all chlorinated benzenes.

5.9. PARTITIONING METHODS

Partitioning methods are employed primarily to separate lipids and oils from environmental pollutants and pesticides. This technique therefore becomes an extraction method as well for organic pollutants from fish tissues, oils, and some food products.

The most commonly used solvent system consists of hexane or petroleum ether with acetonitrile[46] or hexane and N,N-dimethylformamide. These methods provide reliable recoveries for most chlorinated compounds, organophosphates, and esters of chlorinated phenoxyalkanoic acids.

A 3–5-g sample of a fatty tissue is transferred into a 125-ml separatory funnel with petroleum ether to make a volume of 15 ml. Thereafter, 30 ml of petroleum ether-saturated acetonitrile is added, and the mixture is shaken vigorously for 1 min. The two layers are allowed to separate, where the acetonitrile layer will be the bottom phase. The second 1-l separatory funnel is filled with 700 ml of a 2% NaCl solution in distilled water and 100 ml of petroleum ether. The bottom acetonitrile layer from the 125-ml separatory funnel is drained into the 1-l funnel without draining any of the petroleum ether. The remaining petroleum ether is extracted three more times with 30 ml of acetonitrile saturated with petroleum ether. The pressure is released and the content is mixed by turning the funnel upside down several times to assure good recovery. The bottom acetonitrile water layer is drained into a second 1-l separatory funnel. Again, 100 ml of petroleum ether is added to the second separatory funnel and is shaken vigorously. After separation of the layers, the bottom acetonitrile–water layer is drained and discarded. The petroleum ether layer extracts are combined, washed with 100 ml of 2% NaCl in water, and the water layer is discarded. The whole extraction with water is repeated two times. The petroleum ether is drained through a 5–8-cm solution of anhydrous sodium sulfate. The content is placed into a 500-ml Kuderna–Danish concentrator and the combined petroleum ether extracts and wash is evaporated to about 10 ml. A Florisil column cleanup follows.

5.10. MISCELLANEOUS METHODS

Many other highly ingenious methods have been employed for the separation of the organic pollutants. One of the techniques is freeze-drying. Slow freezing, with constant stirring, results in a concentration

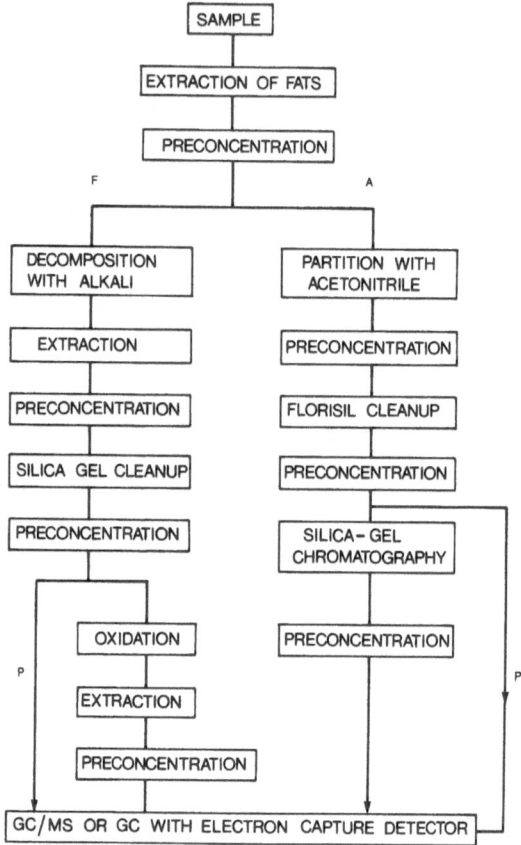

FIGURE 5.8. General scheme for analysis of polychlorinated biphenyls. (A) Alternate methodology; (F) final analysis; (P) preliminary screening.

of organic impurities in the remaining solution. The technique is most effective in water of low salinity. It has been applied to lake water with some success.[47]

The details of a sample preparation procedures for organic environmental contaminants are illustrated in flowcharts of PCB analysis from different matrices shown in Figs. 5.8–5.18. The procedures shown in these flowsheets are applicable not only to PCB but to a variety of organic contaminants with a minimum of modification required.

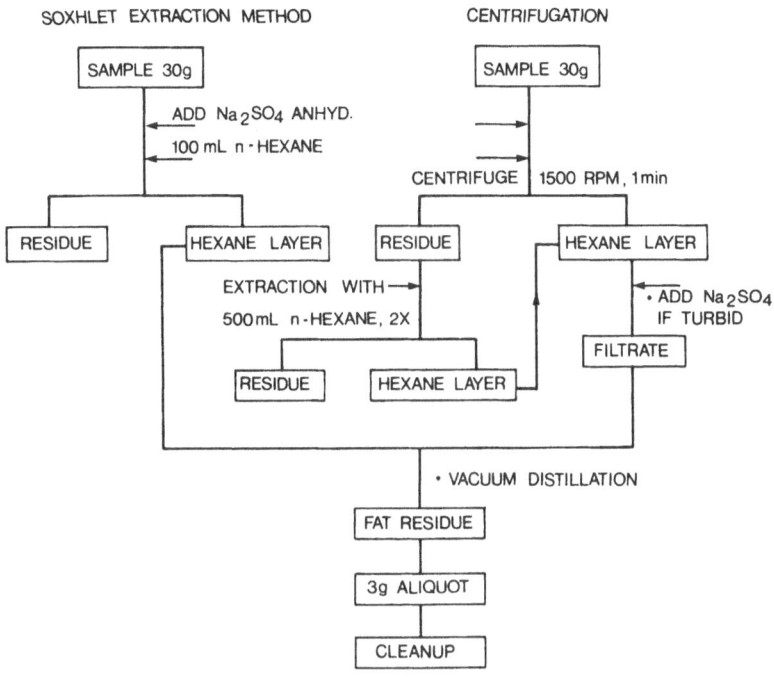

FIGURE 5.9. Extraction scheme for fatty samples using Soxhlet and centrifugation methods.

5.11. AIR SAMPLING

The determination of contaminants in the ambient air is a tremendous task. Most of the existing data concerning the nature of pollution of the ambient atmosphere by organic pollutants has been collected over the period from 1970 to the present. The sampling method utilized is based on impingement in ethylene glycol, and it is cumbersome to use. During the past several years many new sampling devices were developed.[48] For ambient uncontaminated air, sufficiently large samples must be taken to allow detection and measurement at ultratrace levels (pg/m³). Such sampling should be performed over an entire daily cycle if results are to be representative of the average quantities of the substances normally present in the atmosphere. The high-volume air sampler uses a high-volume pump that draws air through a glass fiber filter to collect particulate matter and a solid adsorbent cartridge to trap air pollutants

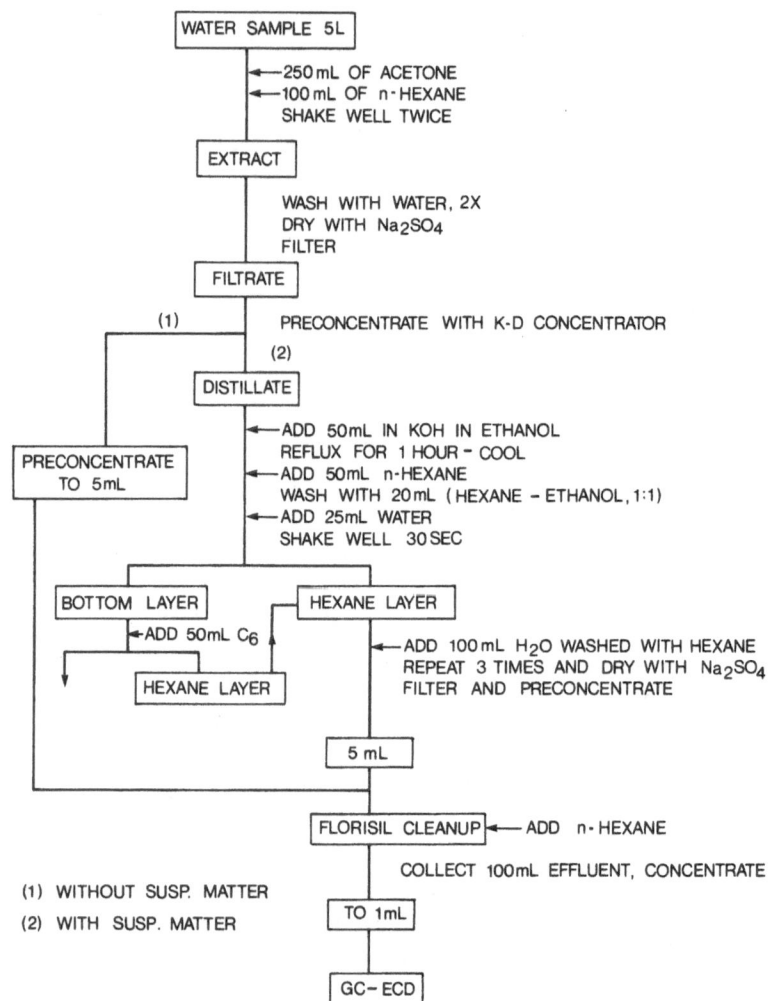

FIGURE 5.10. Extraction of PCB from water (1) without suspended solids; (2) with suspended solids

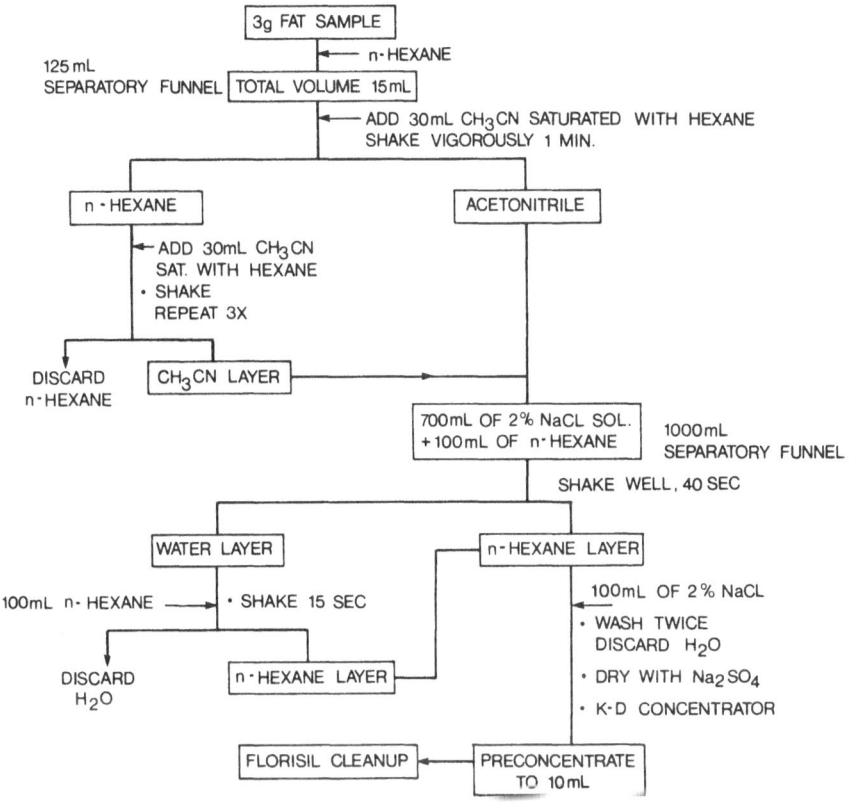

FIGURE 5.11. Acetonitrile–*n*-hexane partitioning scheme.

at sampling rates up to 280 l/min. The sampler is used with a wide variety of adsorbents and has been demonstrated to efficiently collect a number of organochlorine, organophosphate, and carbamate pesticides and herbicides. Calibration of air sampler is performed by means of the Venturi. Two types of sampling adsorbents are recommended as porous microreticular sorbents, e.g., Porapak R (50–80 mesh), Chromosorb 102 (20–40 mesh), Tenax (60–80 mesh), XAD-2. Approximately 30 g of the adsorbent is recommended. Polyurethane foam, cylindrical plugs having 5.5-cm o.d. are used. The polyurethane must be pre-extracted with hexane-diethylether (5%) for 18 hr in a Soxhlet extractor. Both retention and collection efficiency should be established for each resin.

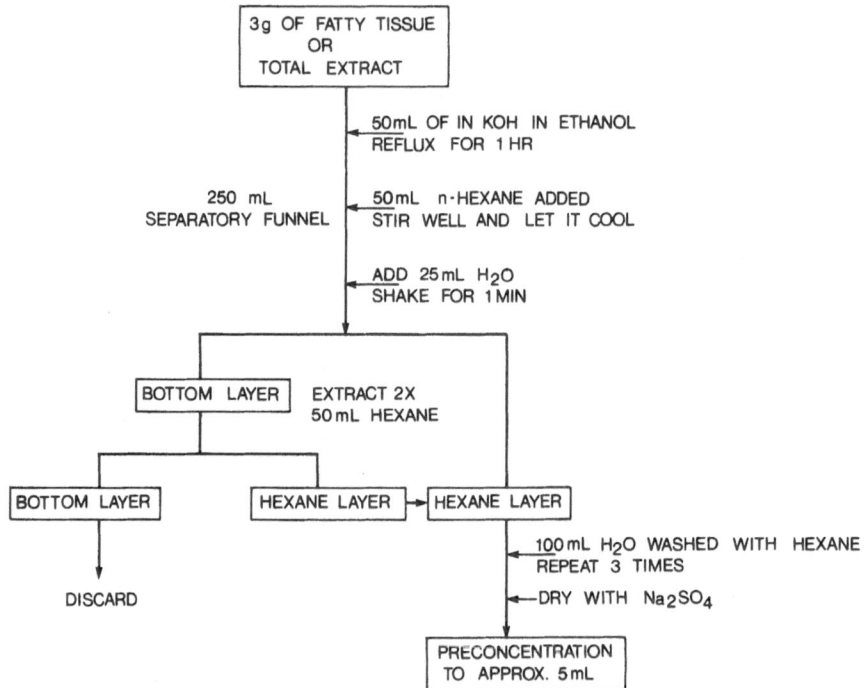

FIGURE 5.12. Decomposition with alcoholic KOH and cleanup procedure.

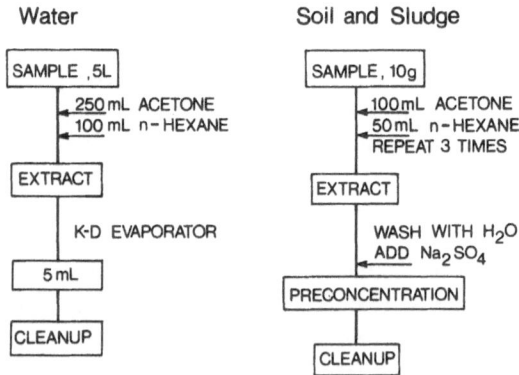

FIGURE 5.13. Sample extraction flowcharts of water, soil, and sludge samples.

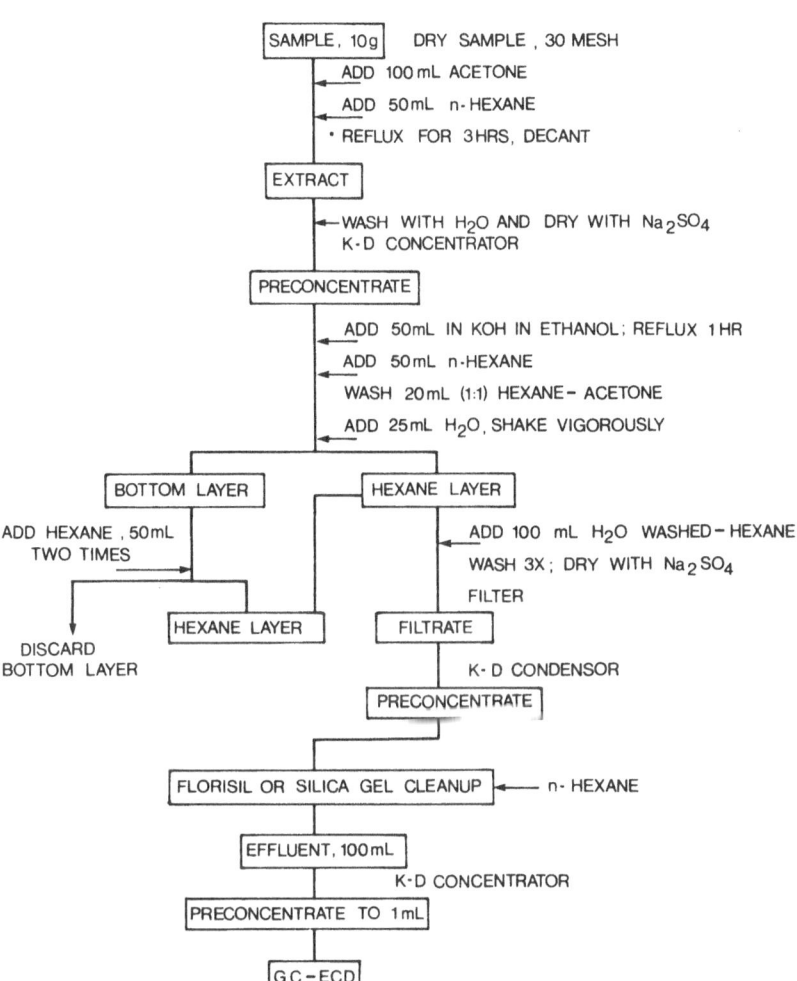

FIGURE 5.14. Extraction, cleanup, and analysis of sediment, soil, and sludge.

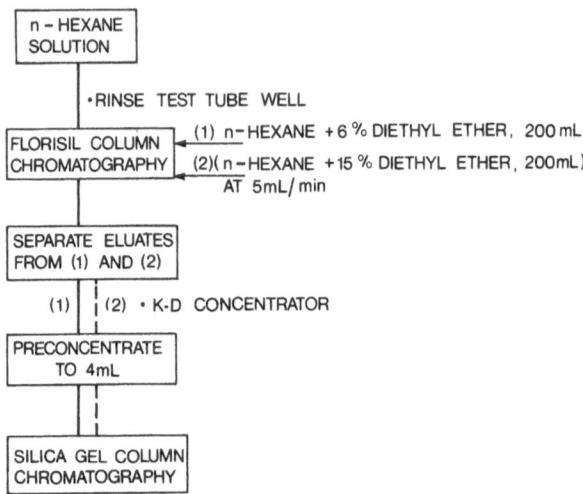

FIGURE 5.15. Florisil column cleanup procedure.

Retention efficiency indicates the capability of the sampler to trap and retain the adsorbed pollutants. It is performed by multiple injections of μl volumes of the pollutant in hexane directly into the sorbent trap. After drying period (1 hr), the fortified trap is placed in front of a second trap in the sampling system. Air is pumped through the system (200 1/ min for 24 hr) to determine breakthrough to the second trap.

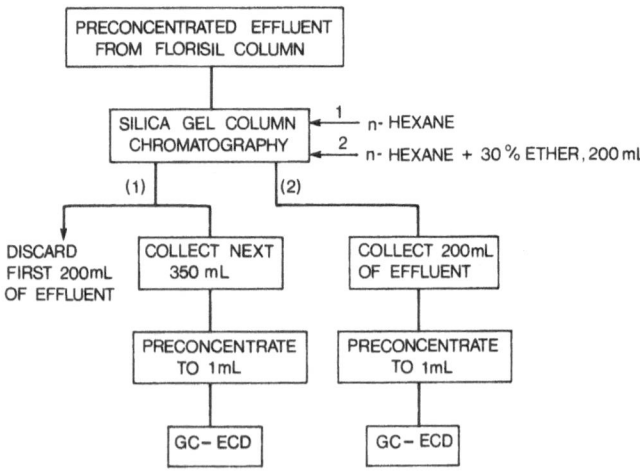

FIGURE 5.16. Silica gel cleanup when pesticides are present.

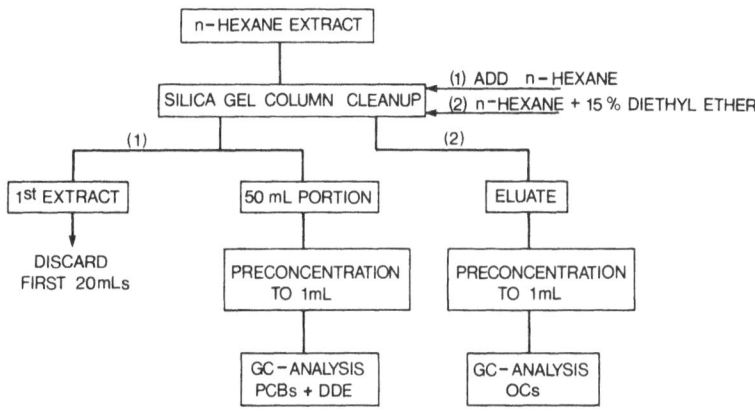

FIGURE 5.17. Silica gel column cleanup procedure.

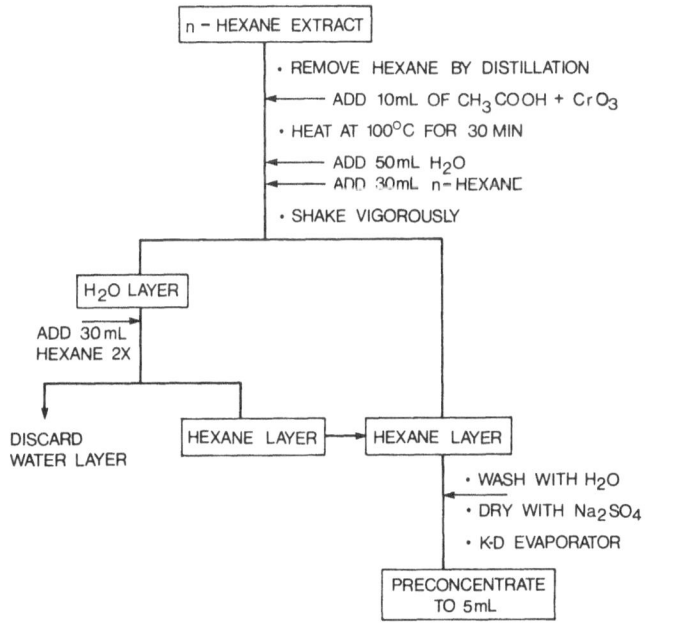

FIGURE 5.18. Oxidation of sample using chromium (VI) oxide.

Collection efficiency is determined by vaporizing individual pollutants or mixture into the intake of the air sampler. A sampling efficiency of 80% is generally considered satisfactory for a collection trap.

The greatest value of a high-volume sampling system is that it provides a large sampling volume of air. The volume should be at least 300 $m^3/24$ hr. In this case sufficient quantities can be collected to detect very low air concentration. For efficiently collected pollutants, detection limits are at the subpicogram (less than 10^{-12} g) per cubic meter of air level.

5.12. SOURCE SAMPLING PROCEDURES

Industrial, contaminated atmospheres generally require less complex sampling procedures because of the higher levels of pollutants present. However, source sampling requires special sampling equipment commensurate with specific sampling needs. Monitoring atmospheres inside dwellings and/or work places requires a sampler that is efficient and quiet.[49,50]

One such type of sampler is manufactured by Environmental Research Corporation (ERCO-Sampler) in St. Paul, Minnesota. Air is drawn at flow rates up to 180 l/min through either or both of two parallel composite filters packed with Porapack R sandwiched between two layers of glass fiber mat. It provides a large sample size with short sampling times. Low-volume indoor samplers are battery-operated pumps containing plugs of polyurethane foam or up to 10 ml of granular adsorbents.

REFERENCES

1. D. McDougall and W. B. Crummett, *Anal. Chem.* **52**, 2242–2249 (1980).
2. C. Pries and H. Compaan, *Chemosphere* **10**(6), 555–559 (1981).
3. H. H. Ku, *Precision Measurement and Calibration*, National Bureau of Standards Spec. Publication No. 300, Vol. 1.
4. R. E. Walpole and R. Myers, *Probability and Statistics for Engineers and Scientists*, Macmillan, New York, 1972, p. 190.
5. F. C. McElroy, T. D. Searl, and R. A. Brown, *Trace Organic Analysis. A New Frontier in Analytical Chemistry*, NBS Spec. Publ. 519, Proceedings of the 9th Materials Research Symposium, Gaithsburg, MD, 1978.
6. J. R. DeVoe (ed.), *American Chem. Society Symposium Ser.* **63** (1977).
7. G. A. Junk, J. J. Richard, M. D. Grieser, D. Witiak, J. L. Witiak, M. D. Arguello, R. Vick, H. J. Svec, J. S. Fritz, and G. Calder, *J. Chromatogr.* **99**, 745–762 (1974).
8. J. S. Fritz, *Acc. Chem. Res.* **10**, 67 (1977).
9. R. Shinohara, M. Koga, F. Shinohara, and T. Hori, *Bunseki Kagaku* **28**, 856 (1977).
10. *Analytical Methods Manual 1979*, Inland Waters Directorate, Water Quality Branch, Environment Canada, Part 5, Naquadat 18165.

11. P. Mills, *J. Assoc. Offic. Anal. Chemists* **51**, 29–32 (1968).

12. *Manual of Analytical Methods for the Analysis of Pesticides in Humans and Environmental Samples*, U.S. Environmental Protection Agency EPA-600/8-80-038, June, 1980.

13. A. V. Kiselev and J. J. Jashin, *Gas-Adsorption Chromatography*, Nauka, Moscow (1967).

14. S. F. Stepan and J. F. Smith, *Water Res.* **11**, 339–342 (1977).

15. A. Tateda and J. S. Fritz, *J. Chromatogr.* **152**, 329–340 (1978).

16. K. Sakodynskii, L. Panina, and N. Klinskaja, *Chromtographia* **7**, 339–344 (1974).

17. V. Leoni, G. Puccetti, and A. Grella, *J. Chromatogr.* **106**, 119–124 (1975).

18. V. Leoni, G. Puccetti, R. J. Colombo, and A. M. D'Ovidio, *J. Chromatogr.* **125**, 399–407 (1976).

19. B. Versino, M. deGroot, and F. Geiss, *Chromatographia* **7**, 302–304 (1974).

20. J. P. Mieue and M. W. Dietrich, *J. Chromatogr. Sci.* **11**, 559–570 (1973).

21. J. J. M. Brown, *J. Chem. Soc. (Ser. A)*, p. 1062 (1970).

22. J. F. Uthe, J. Reincke, and H. Gesser, *Environ. Lett.* **3**, 117 (1972).

23. K. M. Gough and H. D. Gesser, *J. Chromatogr.* **115**, 383–390 (1975).

24. J. D. Navratil, R. E. Sievers, and H. F. Walton, *Anal. Chem.* **49**, 2260–2263 (1977).

25. J. P. Mieue and M. W. Dietrich, *J. Chromatogr. Sci.* **11**, 559–570 (1973).

26. K. Grob, K. Grob, Jr., and G. Grob, *J. Chromatogr.* **106**, 299–315 (1975).

27. D. A. J. Murray, *J. Chromatogr.* **177**, 135–140 (1979).

28. M. Ahnoff and G. Josefsson, *Anal. Chem.* **48**, 1268–1269 (1976).

29. K. Ogan, E. Katz, and W. Slavin, *J. Chromatogr. Sci.* **16**, 517–522 (1978).

30. J. F. K. Huber and R. R. Becker, *J. Chromatogr.* **142**, 765–776 (1977).

31. J. Drozd and J. Novak, *J. Chromatogr.* **165**, 141–165 (1979).

32. J. Janak, in *Chromatography* (E. Heffman, ed.), Reinhold, New York, 1975.

33. K. L. E. Kaiser and B. G. Oliver, *Anal. Chem.* **48**, 2207–2209 (1976).

34. W. F. Cowen, W. J. Cooper, and J. W. Highfill, *Anal. Chem.* **47**, 2483–2485 (1975).

35. C. McAuliffe, *Chem. Technol.*, p. 46 (1971).

36. J. Drozd, J. Novak, and J. A. Rijks, *J. Chromatogr.* **158**, 471–482 (1978).

37. J. Drozd and J. Novak, *J. Chromatogr.* **152**, 55–61 (1978).

38. K. Grob and F. Zuercher, *J. Chromatogr.* **117**, 285–294 (1976).

39. W. D. Snyder, Hewlett-Packard Techn. Paper GC-71, 1978.

40. S. P. Wasik, *J. Chromatogr. Sci.* **12**, 845–848 (1974).

41. R. A. Hites, LABCON 1981, Conference Cassettes, Session TS9A, Professional Cassette Center, Pasadena, CA.

42. T. K. Nielson and S. Kryger, *Dansk Tidskr. Farm.* **43**, 39 (1969).

43. T. J. Siek and R. C. Lindsay, *J. Dairy Sci.* **58**, 1887 (1968).

44. G. D. Veith and L. M. Kiwus, *Bull. Environ. Contam. Technol.* **17**, 631–636 (1977).

45. F. I. Onuska, R. Thomson, and K. Terry, NWRI Report.

46. *Manual for Analytical Quality Control for Pesticides and Related Compounds in Human and Environmental Samples*, Environmental Protection Agency EPA-600, 1-79-008, Research Triangle Park, NC, January, 1979, p. 240.

47. P. A. Kammerer, Jr., and G. F. Lee, *Environ. Sci. Technol.* **3**, 276–278 (1969).

48. R. G. Lewis, K. E. MacLeod, and M. D. Jackson, Proceedings of the 2nd Chemical Congress, ACS-Chem. Soc. of Japan, Paper No. 65, Honolulu, Hawaii, April, 1979.

49. E. Wanters, P. Sandra, and M. Verzele, *J. Chromatogr.* **170**, 125–131 (1979).

50. H. H. Hill, Jr., K. W. Chan, and F. W. Karasek, *J. Chromatogr.* **131**, 245–252 (1977).

CHAPTER SIX

THE RETENTION INDEX SYSTEM

6.1. INTRODUCTION

The retention behavior of a compound contains the qualitative information available from gas chromatography. This has long been recognized and stimulated many efforts to establish useful concepts for developing systematic data broadly applicable to analytical problems. The concept of specific retention volume was suggested to make retention data essentially independent of the variable chromatographic parameters.[1] Its calculation of the volume of carrier gas reduced to 0 °C required to elute a compound takes into account the column temperature, carrier gas flow rate, column pressure drop, and the amount of liquid phase present. Although it is theoretically correct, specific retention volume is not used in practice because it is awkward to calculate, does not allow for retention changes with temperature, and calls for values that may change. The use of relative retention data is at present the most accepted method. In the determination of relative retention, the retention of both a reference compound and the sample components are determined under identical chromatographic conditions. Since it is almost impossible to establish only one compound to be a reference, a more universal method is needed for wide usage. It is known that under isothermal conditions a plot of the logarithm of the adjusted retention time t_R' versus molecular weight of a homologous series is linear. This behavior forms the basis of the retention index systems now in use. Essentially, a series of reference compounds are used to establish a linear reference scale into which the retention of a compound will fall.

6.2. KOVATS RETENTION INDEX SYSTEM

In 1958 Kovats proposed a retention index system in which the series of compounds used to establish the linear reference scale are n-hydrocarbons.[2,3] This system is defined by

$$I_x = 100 \cdot \left[\frac{\log t_x' - \log t_n'}{\log t_{n+1}' - \log t_n'} \right] + 100n \qquad (6.1)$$

where I_x is the retention index of compound x, n is the number of carbon atoms in the n-alkane reference, t_n' is the adjusted retention time of the n-alkane reference with n carbon atoms, and t_{n+1}' is the adjusted retention time of the n-alkane reference with $n + 1$ carbon atoms.

The data is obtained under isothermal conditions. In his original equation Kovats used the specific retention volume V_g, but it is obvious this term can be replaced by the term of adjusted retention time, t_x', which is much easier to measure directly from the chromatogram. More precise and easier to determine results can be obtained if it is possible to use a mixture of the n-alkane standards as an internal standard rather than an external standard.

6.2.1. Calculation and Precision

Figure 6.1 illustrates the retention index system and a graphical interpolation means of determining the I_x value of an unknown. To obtain precise results that can be repeated between laboratories, certain sources of error need to be minimized. Instrumental sources arise from inaccuracies in column temperature and carrier gas flow rate between runs if the external standard method is used. Since each stationary phase gives

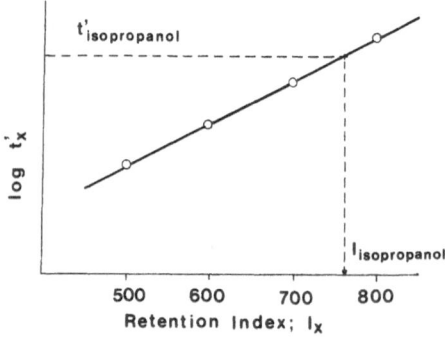

FIGURE 6.1. Determination of the Kovats retention index. The logarithms of the adjusted retention times of the bracketing n-alkanes are plotted (ordinate) as functions of their retention indices (abscissa), and a line is constructed through these points. Where the logarithm of the adjusted retention time of isopropanol intersects that line, a perpendicular is dropped to the abscissa that determines the retention index of the isopropanol.

a characteristic I_x value, its purity must be well defined. Changes in the properties of a phase such as oxidation by traces of oxygen in the carrier gas will significantly alter the I_x value. The I_x value of a compound will be altered by lack of resolution of the compound from one coeluting so that the peak maximum is shifted. Use of small samples will minimize obtaining distorted peak shapes whose maxima are difficult to determine accurately. Active, uncovered sites on the column wall will contribute significant variation to the I_x values, particularly for polar compounds.

Kovats indicated that a reproducibility of $2I_x$ units was within experimental error for his packed-column work. With carefully prepared WCOT columns more precise data is possible. Rijks has made an intensive study of the parameters that influence I_x values.[4] Using high-resolution WCOT columns and very precise control of carrier gas flow and temperature, his results indicated errors can be held to less than 0.02%, even between laboratories.

Modern, computerized gas chromatographs are designed to facilitate determination of very precise I_x values. Computerization permits the design of the analyzer to give accurate temperature and carrier gas flow for WCOT columns. The computer uses peak maxima values electronically measured to automatically calculate I_x values for a sample. An example of such data is seen in Fig. 6.2.

6.2.2. Effects of Temperature

The effect of temperature on I_x depends upon the relative changes in partition properties between the reference compounds and the compounds being analyzed. This change is not too great for most compounds on most stationary phases, except for some high molecular weight, highly polar compounds. The I_x values increase according to a hyperbolic curve as temperature increases.[3] Within a 50 °C range, this change is expressed by a linear function,

$$\frac{\Delta I}{10°} = 10 \left[\frac{I_{T_2} - I_{T_1}}{(T2 - T1)} \right] \tag{6.2}$$

where $\Delta I/10°$ is the incremental change in I_x for 10 °C, T_1 is the lower, and T_2 is the higher temperature. The $\Delta I/10°$ value for most compounds ranges from two to six I_x units, although values in excess of 30 may be experienced.

Programmed temperature gas chromatography is widely practiced, especially when analyzing complex mixtures on WCOT columns. Van

```
300 DATA 7.120,23.112,2,0.8,1,19

START PRGM
```

NO	RETENTION TIME	RETENTION INDEX	AREA	AREA (/UL)	PEAK WIDTH
--	---------	---------	----	----	-----
1	2.175	176.443	71.02	88.77	0.000
2	2.335	177.206	159.37	199.21	.035
3	2.519	178.082	261.27	326.39	.027
4	2.616	178.544	327.52	409.40	.027
5	2.804	179.440	879.33	1099.16	.028
6	2.870	179.754	56.04	70.05	0.000
7	3.095	180.826	52.58	65.73	0.000
8	3.132	181.002	33.17	41.46	0.000
9	3.597	183.217	38.04	47.55	.040
10	4.370	186.900	25.15	31.44	.050
11	7.170	200.238	77.38	96.73	.060
12	7.320	200.955	16.27	20.33	0.000
13	8.157	204.940	86.06	107.58	.064
14	8.531	206.722	21.83	27.29	0.000
15	8.963	208.780	75.86	94.83	.058
16	9.946	213.462	41.31	51.64	.097
17	13.074	228.363	17.13	21.41	0.000
18	17.417	249.052	16.56	20.70	0.000
19	25.880	289.367	124.93	156.16	.118

```
END OF PROGRAM
```

FIGURE 6.2. Printout obtained for WCOT analysis with HP-5880 computerized GC showing calculated retention index values.

den Dool and Kratz showed that for programmed temperature operation, log t_x' in the Kovats equation can be replaced with t_x to give an approximately linear index scale.[5] The equation then becomes:

$$I_x = 100\left[\frac{t_x - t_n}{t_{n+1} - t_n}\right] + 100n \tag{6.3}$$

There seems to be a correlation between I_x values obtained with programmed temperature operation and those from isothermal operation. Guiochon found for a series of compounds that the I_x programmed temperature value corresponded to the I_x isothermal value obtained at a temperature of 20 °C lower.[6] On a theoretical basis Giddings showed that a programmed temperature chromatographic separation corresponds well to an isothermal one at a temperature T' given by 0.92 retention temperature T_R to a good approximation.[7] In any event, I_x values obtained

with programmed temperature operation are best used in connection
with analyses in which the programmed conditions are repeated exactly
each time.

6.3. COMPOUND IDENTIFICATION BY RETENTION INDEX

Before the widespread use of GC/MS systems in which the mass
spectrum of the compound in a GC peak provided the identification data,
there were many attempts to systemize I_x data in the literature to provide
readily available tabulated data for identification.[8] Even with high-resolution WCOT columns and precise I_x determination, the possibility of
two compounds exhibiting identical retention behavior on a given column
exists. By determining the I_x values on two high-resolution columns of
different polarity, the certainty of identification is increased. For example,
both tetrahydrofuran and 1-butanol give an I_x value of 618 on Apiezon
L stationary phase, while on Emulphor 0 the corresponding I_x values are
800 and 973. It is now generally accepted that such data constitute a
reasonably positive identification. The plot in Fig. 6.3 not only illustrates
this point but also shows the correlations and linearities with functional
groups and molecular weights inherent in the I_x system.

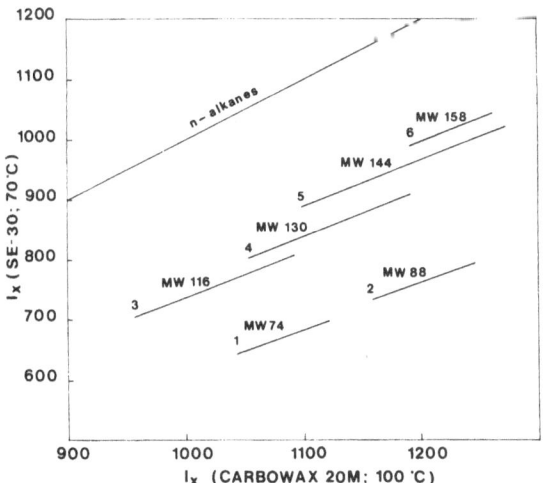

FIGURE 6.3. A plot of the retention index on Carbowax-20M versus that for SE-30 for
alcohols (1,2) and esters (3–6) relative to functional group and molecular weight, and the
similarity of slope exhibited by groups with similar functional groups.

6.4. OTHER RETENTION SYSTEMS

Use of the Kovats retention index system with its n-hydrocarbon reference standards is not always feasible, especially when applied to polar compounds on polar stationary phases because it requires the use of very high molecular weight hydrocarbons. A number of systems completely analogous to the Kovats system except for the use of a different series of reference compounds such as saturated fatty acids, n-ethyl esters, and methyl ketones have been suggested.[9]

The analysis of complex mixtures of polycylic aromatic hydrocarbons (PAH) using programmed temperature conditions illustrates the value of establishing an I_x system based on a series of PAH internal standards.[10,11] A new I_x system is defined based on the series of PAH internal standards of naphthalene, phenanthrene, chrysene, and picene. The analysis of complex mixtures of PAH compounds on WCOT columns gives separation of over 150 compounds. The large number of possible isomers and the unavailability of reference compounds for each isomer makes availability of good, standardized retention data essential. The problems of poor reproducibility and need for high molecular weight n-alkanes in the Kovats system are alievated by using the above four PAH compounds to establish the reference framework. Table 6.1 illustrates the improvement in the data possible. In the determination of I_x values of over 200 PAH compounds using this system, the 95% confidence limits for four

TABLE 6.1

Kovats Retention Indices (RI) and PAH Indices from WCOT Columns with a Different Stationary Phase Film Thickness of SE-52

Compound	Kovats retention index Column		PAH index Column	
	A	B	A	B
Naphthalene	1168.74	1166.21	200.00	200.00
Acenaphthylene	1425.03	1413.00	244.65	244.67
Fluorene	1555.87	1549.28	268.14	268.22
Phenanthrene	1744.70	1734.95	300.00	300.00
Anthracene	1754.20	1744.40	301.73	301.76
Pyrene	2063.99	2048.56	351.13	351.25
Benzo[a]fluorene	2167.27	2153.04	366.64	366.75
Benzo[e]pyrene	2770.65	2751.03	450.66	450.80
Perylene	2812.49	2815.42	456.12	456.23

measurements on each PAH are $\pm 0.25 I_x$ units.[11] The equation used for these temperature-programmed data is:

$$I_x = 100\left[\frac{t_x - t_z}{t_{z+1} - t_z}\right] + 100z \qquad (6.4)$$

where t_x is the retention of the PAH compound and t_z and t_{z+1} are the retention times of the bracketing reference compounds with z aromatic rings.

The value of the I_x data used in connection with mass spectra obtained by GC/MS systems in mixtures containing multiple isomeric sets of PAH comounds is clear because the mass spectra of the isomers are so similar that the retention data must be used to differentiate them. The assignment of Kovats I_x values automatically in computerized GC/MS systems has been described by Nau.[12] This greatly facilitates generation of the data.

REFERENCES

1. A. B. Littlewood, C. S. G. Phillips, and D. T. Price, *J. Chem. Soc. (London)*, 1480–1483 (1955).
2. E. Kovats, *Helv. Chim. Acta* **41**, 1915–1932 (1958).
3. L. S. Ettre, *Anal. Chem.* **36**, 31A–41A (1964).
4. J. A. Rijks, "Characterization of Hydrocarbons by Gas Chromatography. Means of Improving Accuracy," Doctoral thesis, Technical University, Eindhoven, The Netherlands, 1973.
5. H. Van den Dool and P. D. Kratz, *J. Chromatogr.* **11**, 463–469 (1963).
6. G. Guiochon, *Anal. Chem.* **36**, 661–663 (1964).
7. J. C. Giddings, *Gas Chromatography* (N. Brenner, J. E. Callen, and M. D. Weiss, eds.), Academic, New York, 1962, p. 57.
8. O. E. Schupp and J. S. Lewis, *Res. Dev.* **21**, 24–28 (1970).
9. T. K. Miwa, K. L. Micolajczak, F. R. Earle, and I. A. Wolf, *Anal. Chem.* **32**, 1739–1743 (1960).
10. M. L. Lee, D. L. Vassilaros, C. M. White, and M. Novotny, *Anal. Chem.* **51**, 768–774 (1979).
11. D. L. Vassilaros, R. C. Kong, D. W. Later, and M. L. Lee, *J. Chromatogr.* **252**, 1–20 (1982).
12. H. Nau and K. Biemann, *Anal. Chem.* **46**, 426–431 (1974).

CHAPTER SEVEN

QUANTITATIVE ANALYSIS

7.1. GENERAL CONSIDERATIONS

The basic principles and conditions of gas chromatography necessary for quantitative gas chromatographic analysis have been described by Novak.[1] Quantitative GC analysis is always based on the previous identification of the compounds involved. In complex environment mixtures, GC analysis is preceded by a cleanup procedure to remove bulk matrix compounds that would interfere with the analysis. In this way a reliable quantitative determination at the microgram to low picogram level is possible. Quantitative analysis with GC detectors places special demands on the measurement principle used for the detector. The signal intensity depends not only on the amount of the sample but also on a number of experimental parameters of the detection system. In practice a comparative measurement of signal represented by the peak area or a peak height of the unknown substance, unknown concentration, and a standard of known concentration can be performed in several ways:

1. A calibration curve is drawn up of the signal as a function of the amount of substance. The unknown concentration is determined from the intensity of its signal on the standard calibration curve. Here, quantitative determination requires a comparison of two separate measurements, such as an unknown concentration and a known calibration mixture which is used as an external standard. The standard addition method may also be used. After the sample has been divided into two equal parts, a known amount of the compound to be determined quantitatively is added to one of them. The signal intensities of the untreated sample and of the sample with the standard are then compared.

2. If a reference compound in known amount is added directly to the sample to be analyzed and is analyzed together with the compound of the sample, the method of the quantitative determination is based on an internal standard. Homologues, or chemically similar compounds, may serve as internal standards. Before quantitative analysis, the suitability of the selected internal standard must be checked and its response factor must be calculated.

GC quantitation involves the comparison of standards and samples, their areas or peak heights, retention times, and retention temperatures for temperature-programmed operation.

Two requirements necessary for quantitative GC analysis are having a reproducible chromatographic procedure and also a linear response of the detection system for the particular compounds of interest. The quantitation procedure can be accomplished using either a peak height measurement or a peak area measurement.

7.1.1. Peak Height Measurements

Measurements are made for each peak of interest, from the baseline to the top of the peak. The drawbacks to usage of peak heights are directly related to the effects of chromatographic conditions, such as flow rate and elution temperature, on peak height sensitivity. Errors may also arise from poor measurements due to insufficient resolution, drifting baseline, and peaks which are very small.

7.1.2. Peak Area Measurements

Many methods have been employed for measurements of peak areas.[2-4] Most of these—planimetry, cut-and-weight, triangulation, and mechanical disc and ball integrator—are of historical interest only. Beginning in 1977, electronic digital integrating has made all others obsolete. Not only is digital integration extremely accurate, but the data is generated in a form suitable for the calculation of other analytical parameters.

7.1.3. Digital Integrators

Computing integrators have been developed that perform baseline and peak detection, baseline correction, area integration, and allocation of fused peaks. The integrator will compute the appropriate response factors, placing them into memory for future calculations. These systems provide the fastest and the most accurate values for quantitation of very narrow WCOT column peaks.

An ideal chromatographic peak starts at zero signal, rises to a maximum, and returns to zero. The baseline does not change. However, in practice, the baseline can drift in both a positive or a negative direction. If integration is based on a horizontal baseline, inaccurate results would be obtained. Some systems overcome this problem with an offset voltage. The integrator establishes a baseline and rejects area that is not related to the peak being integrated. Integration is performed continuously and can be transferred to the processor every 100 msec. During the integration process, the computing integrator retains a 500-msec value in memory, which is represented by five data bunches. If the baseline does not change, the first value (first 100 msec) is discarded. When a peak emerges from the column, it is detected as a slope rise, and the integrator retains the area accumulated in the five preceding bunches and then continues to accumulate the area throughout the peak. During the elution of a GC peak, the slope of an integration signal begins as a positive slope from the baseline, then decreases in magnitude, and becomes zero when the chromatographic peak reaches a maximum. The slope then becomes negative, increases negatively, and returns to zero when the chromatographic peak returns to baseline.

7.1.4. Microcomputers

The microprocessor has given rise to a growing trend in distribution of processing power. The small size of a central processing unit (CPU), which still retains moderate processing power, means that a microcomputer can be produced using only a few integrated circuits. The genesis of moderately priced minicomputers brought processing into the GC laboratory.[5] When input–output devices are able to communicate with the gas chromatograph as well as with the analyst, interactive operations can be performed. This is shown in Fig. 7.1.

The analyst interacts with the minicomputer, which in turn collects the data, makes decisions, calculates variables, and presents results to the printer–plotter during the gas chromatographic run.

Computerization of GC instrumentation has advanced to the point that the gas chromatograph is designed around the minicomputer. The operator directs the operation of the computer, which in turn can prompt the operator to enter and check all variables.

The microcomputer performs data acquisition and data reduction in much the same way as a computing integrator. The same sort of commands are used to instruct the microcomputer for data handling. The significant performance difference between the two devices is that the

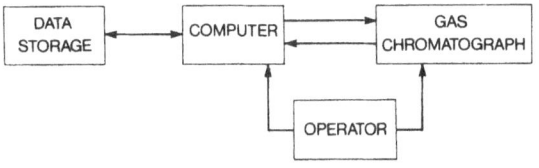

FIGURE 7.1. The interactive operational diagram of a microcomputer system.

microcomputer can process the data and perform calculations much faster than the computing integrator. Since the computer has a user-accessible language built in, an operator can modify or write programs for statistical evaluation, and print tables of GC data from a series of runs.

For a microprocessor-based system to qualify as a microcomputer system, the processor must consist of memory, a terminal, software, mass storage, and input–output devices (Fig. 7.1). Microcomputers also permit control and interfacing of peripheral equipment not possible with other systems.

Once the gas chromatographic peaks have been quantified, the operator has numbers representing peak areas or peak heights. These numbers may be converted to real or absolute concentrations using various methods employing such different mathematical manipulations as normalization, external standardization, internal standardization, and standard addition.

7.2. PRINCIPLES OF QUANTITATIVE ANALYSIS

7.2.1. Normalization

This method assumes that area percent is equivalent to concentration percent. The basic principles of quantitative analysis by gas chromatography states that area A under a chromatographic peak is a function of analyte concentration C.

$$A = f(C) \qquad (7.1)$$

Ideally, the graphic representation of this relationship is a linear plot in which the slope S represents the proportionality constant between area and concentration:

$$S = \frac{\text{area}}{\text{concentration}} \qquad (7.2)$$

However, when several components are present at the same concentration, the areas under their respective peaks are not necessarily the same. This variation in response is related to detector sensitivity being different for each compound.

Area normalization for compounds with similar or equal detector response can be calculated as follows:

$$\text{Area normalization} = \frac{\text{area of peak } A}{\text{total peak areas}} \times 100 \qquad (7.3)$$

If individual components in the analyte have different responses, area normalization using response factors to correct for variations in detector response must be introduced.

To calculate response factors, a calibration curve with a minimum of three concentrations for each of the components must be made. A plot of data from the chromatogram representing concentrations (weight) against peak area is constructed (Fig. 7.2). In Fig. 7.2 the slopes of each plot represent the response factors of the corresponding components. The corrected area of each component is calculated by dividing its peak area by the response factor.

To simplify calculations, the chromatographer usually performs a calibration step based upon determination of relative detector response employing a reference compound. The calculated value is expressed as the relative response factor (RRF). The relative response factor represents the ratio between the slope of the analyte compound and the slope of the reference component in the mixture. Since the slopes are the proportionality constants, the ratio between proportionality constants gives the relative response factor (Fig. 7.3):

$$\text{RRF}_x = \frac{S_x}{S_y} = \frac{A_x/C_y}{A_y/C_y} \qquad (7.4)$$

where S_y is the proportionality constant of the reference compound Y (see Fig. 7.2).

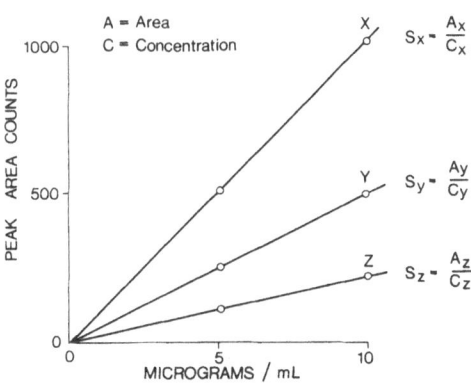

FIGURE 7.2. The calibration plot of area versus concentration for three different components.

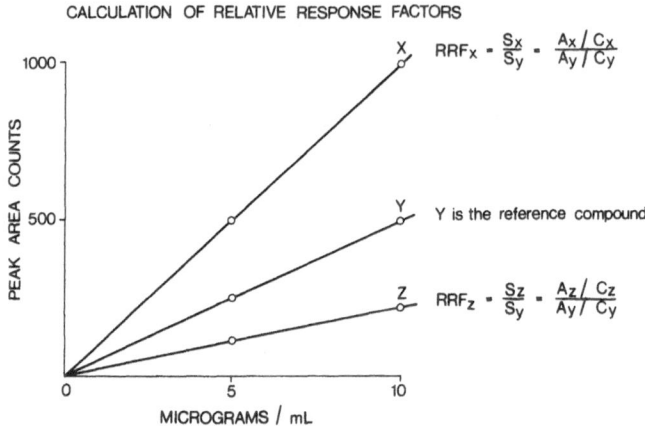

FIGURE 7.3. Plot of the relative response factors as related to the reference compound.

The relative response factors are used to correct the areas of chromatographic peaks. This procedure compensates for the difference in detector response and enables the chromatographer to determine the relationship of corrected peak area to total peak area. This in turn enables him to determine the concentration percent of the components in a mixture.

The corrected area normalization method is employed if the detector response is not the same for all components in the mixture. It consists of two steps. The first step, calibration, determines relative response factors by measuring peak areas for known amounts of pure standards. The second step, analysis, employs relative response factors from the calibration step to obtain corrected peak areas which are directly related to concentration.

Concentration percent (C) then is calculated as follows:

$$C = \frac{A \cdot \text{RRF}}{\sum_{a=1}^{n} A \cdot \text{RRF}} \times 100 \times \text{SF} \tag{7.5}$$

where SF is the scale factor.

The scale factor (SF) is a fraction less than unity used to compensate for any component in a mixture which is not detected but whose concentration is known. This method determines the area under each chromatographic peak, calculates RRFs and applies them to each peak when desired, obtains the total area of all peaks, and calculates individual peak concentration.

7.2.2. External Standardization

The external standardization method is commonly used in environmental analysis. In order to use this method, the linearity of response for the components of interest in the expected concentration range must be verified. Calibration curves of pure standards of varying concentration are first prepared. It is essential to the accuracy of the measurement that injection volumes of the sample and pure standards always be identical. In a determination in which area is a linear function of concentration, the slope S can be expressed as area divided by concentration. This expression is also the relative response factor, RRF, used in calculation of results. This leads to an expression for concentration (C) in terms of area (A) and RRF:

$$C = \frac{A}{\mathrm{RRF}} \tag{7.6}$$

Also, the external standard method is employed when only a few components in a complex mixture are of interest. The chromatographer has to know exactly the concentration of the standard solution and the volume injected. He can increase the accuracy of the measurement of the RRF by repeating the measurement of the analyte many times or by using a number of different concentration values of the same standard to obtain a calibration graph passing through the origin, and having a constant slope. This method is expressed in Eq. (7.7) and is ideal for use with an autosampler.

$$C_x = A_x \frac{C_s^{\,\circ}}{A_s} = \frac{A_x}{S_x} \tag{7.7}$$

where C_x is the concentration of compound, A_x is the area of component, C_s is the concentration of standard, A_s is the standard, and S_x is the slope.

The accuracy of the external standard method is directly related to the accuracy with which a quantity of the analyte in the standard is known. It is clear the above equation is valid if the response represents a linear relationship. When the response is not linear, a calibration graph of area versus concentration must be constructed. This quantitation assumes that all conditions are equal, or at least sufficiently similar, and that the area of the unknown analyte A_x will fall on the graph in the expected concentration range. In the case of injecting a sample containing more than one component, a number of equations are obtained one for each of the components. The ratio of the weights between, e.g., component 2 and 3 is $R_c = C_2/C_3$. This, however, does not mean that the ratio of the corresponding peak areas, $R_A = A_2/A_3$, is the same. Since the detector will, in general, respond differently to each substance, the ratios R_c and R_A will be different. In fact, R_A/R_c represents the sensitivity factor of component 2 relative to component 3 in the mixture, under the given conditions. When the response is not linear for both components in the mixture, the relative sensitivity factor R_A/R_c may not be constant for changing concentrations.

In summary, the external standard method can be set up easily because only substances of interest need to be calibrated. On the other hand it imposes two restrictions that must be taken into account: (a) chromatographic conditions, such as column temperature and the velocity of the carrier gas, must be very stable and (b) sample size must be very carefully controlled.

7.2.3. Internal Standardization

The internal standardization technique requires the addition of a known compound of known concentration to the sample for comparative determination. When applying this method, the ratio of the area of a component peak to the area of the internal standard peak is used for calculating this ratio for each component of interest.[6] This ratio is known as the area ratio (A_x/A_{IS}). Since the internal standard (IS) is of fixed volume and known concentration, compensations for variations in sample injection size and sample manipulation can be made. The compensation is based upon comparison of the peak area of the internal standard with the peak area of the components of interest. Area ratios obtained must be corrected for differences in detector response by applying the appropriate RRF to both component peak area and IS peak area. Using this method of quantitation, it is not necessary for all components present in the solute to be eluted from the column; only those compounds of interest and the internal standard need be eluted. As with the normalization method, a response factor can be calculated by mixing a known concentration of pure internal standard with various concentrations of pure component standards. A chromatogram is then run for each of the standards, and measurements of peak areas are made for both the component of interest and the internal standard peaks. The ratios of concentration of component i to internal standard are shown in Fig. 7.4. After applying the proper RRF values to the plots, all plots coincide with the internal

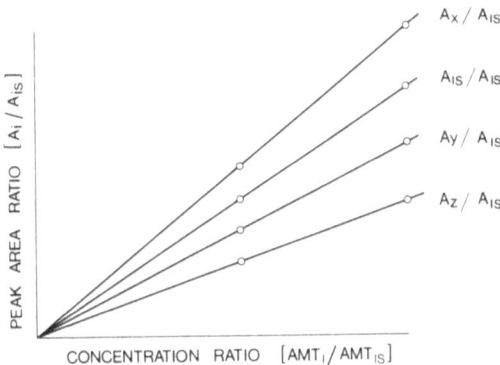

FIGURE 7.4. Internal standardization calibration plot.

standard plot. As a consequence, a particular value of the corrected area ratio will give the same amount (concentration) ratio for all components (Fig. 7.5).

The amount of the component of interest equals the amount ratio of the component multiplied by the amount of internal standard. The product multiplied by 100 and divided by the amount of sample equals the percent of the component.

Concentrations is unknown samples containing components of interest are determined as follows:

$$C_x = \frac{\text{peak area of component}}{\text{peak area of internal standard}} \times RRF \tag{7.8}$$

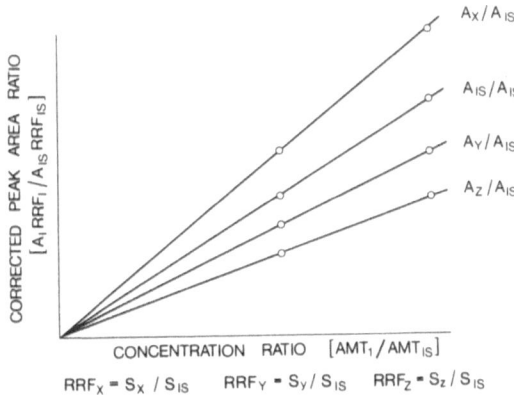

FIGURE 7.5. Illustration of corrected area ratios in the internal standardization method.

An internal standard (IS) must meet the following criteria:

   (i)   never be a component of the solute,

   (ii)   be completely resolved from sample component peaks,

   (iii)   be available in pure form,

   (iv)   be added at a concentration similar or close to that of the components of interest, and

   (v)   not react with any of the sample components.

When using an internal standard, it is recommended that a three-point calibration curve be run routinely to check for degradation of standard and as an instrument check.

A universal calculation procedure applicable for any GC determination where an unknown peak is calculated against a peak resulting from injection of a sample containing a standard of known concentration can be written as

$$C_x = \frac{a_i \cdot A_x \cdot D}{w \cdot A_i} \qquad (7.9)$$

where

$$D = \frac{\text{ml of extracting solvent} \times \text{volume of final extract in } \mu l}{\text{aliquot taken of original extract} \times \mu l \text{ injected}} \qquad (7.10)$$

and C_x is the concentration of unknown component in ppb, a_i is the weight of a component represented by the standard peak in ng, A_x is the area of sample peak, A_i is the area of standard peak, and w is the weight (g) or volume (ml) of original sample.

In calculating the dilution factor (D) for a situation where the final extract concentrate contains the entire original sample, the values for the volume (ml) of extracting solvent and the aliquot taken of original extract would cancel out so that the dilution factor would become

$$D = \frac{\text{volume of final extract in } \mu l}{\mu l \text{ injected}} \qquad (7.11)$$

Environmental analytical data are generally reported as mg/kg, which is equal to ppm units, μg/kg is equal to parts per billion (ppb) units, and ng/kg is equal to parts per trillion. Contaminants in water are commonly expressed as micrograms per liter (μg/l), which is equivalent to ppb.

7.2.4. Standard Addition

While the standard addition method is occasionally employed, in texts on chromatography it is often completely ignored.[1,6,7] This method of quantitation is a combination of internal and external standard methods. It is generally employed when working in a nonlinear portion of the detector response. A mathematical treatment is presented in the literature.[8]

The standard addition method has been employed only recently as a means of quantitation in headspace analysis.[9,10] In this method a volume of headspace gas, V_g, is withdrawn from the equilibrated system and injected into the gas chromatographic column, obtaining a peak area A_i for the component to be determined. Then a defined mass, W_s, of the component being determined (i) is introduced into the GC system. After the latter has re-equilibrated, a volume of the headspace gas, V_g', is again withdrawn and chromatographed to obtain a peak area A_i'. The total mass W_i of component i in the original gas-condensed phase is calculated as

$$W_i = \frac{W_s - W_i}{(A_i'V_g/A_iV_g') - 1} \tag{7.12}$$

Hydrophilic solutes at concentrations in the range of 1–100 mg/l and both aliphatic and aromatic hydrocarbons at concentration of 0.001–20 mg/l in the aqueous phase can be determined with relative error of 20 and 10%, respectively.

7.3. STATISTICAL CONSIDERATIONS

There are specific problems in the reporting of low-level concentrations that are associated with the question of whether a substance is present or not. Since Avogadro's number is very large, it could be argued that one should never claim that a substance is not present. It is necessary to employ some practical measure of the absence of a compound. If measurement is being made in micrograms per liter the presence of a few nanograms per liter is irrelevant.[11]

In environmental analyses a criterion of detection is introduced. It refers to the minimum analytical result that must be observed before it

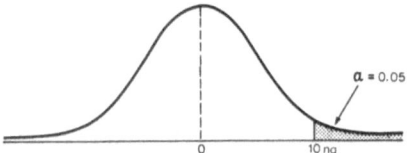

FIGURE 7.6. Definition of the criterion of detection and an α error for finding something which is not there.

can be stated that a substance has been detected with an acceptable probability that the statement is true. For example, suppose that the standard deviation, σ, of an analytical procedure is 6 ng/l and that an α of 0.05 (see Fig. 7.6) is deemed acceptable so that the probability of making an error of finding something which is not there is set at 5%.[12] The criterion of detection can then be found from a table of cumulative normal probabilities to be 1.6456. [1.6456 × 6 ng/l = ng/l.]

Any value observed below 10 ng/l would be reported as less than the criterion of detection since to report such a value otherwise would increase the probability of making this error beyond 5%. Once the criterion of detection has been set, the probability of making an error of not finding a substance which is there, β, or its complement, $1 - \beta$, the probability of detecting the substance when it is present, can be determined for given situations (Fig. 7.7).

Consider the same example as above with a criterion of detection of 10 ng/l. Suppose that the concentration of the sample being analyzed is 10 ng/l and is equal to the criterion of detection. If all analytical results below the criterion of detection were reported as such, than the probability of detecting the sustance would be 50% at 0.5.

From this example it can be seen that the probability of discerning a substance when its concentration is equal to the criterion of detection is hardly overwhelming. In order for the probability of not finding the substance which is there to be equal to the probability of finding a substance which is not there, error β is equal to α, then the concentration of the sample being analyzed must be twice the criterion of detection.

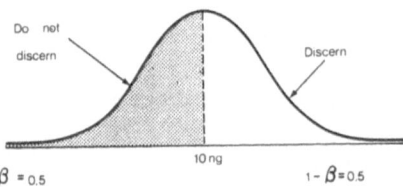

FIGURE 7.7. Definition of a β error representing not finding something which is there.

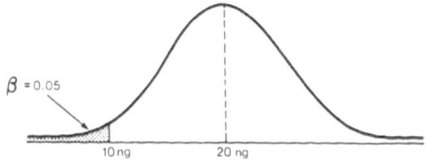

FIGURE 7.8. Limit of detection definition, where α error is equal to β error.

This concentration of twice the criterion of detection is the limit of detection (Fig. 7.8).

It should also be recognized that when the probability of making an error of finding a substance which is not there is decreased by selecting a lower α level, the probability of making an error of not finding something which is there is increased.

The limit of detection is considered within the context of quantification. Therefore, it is defined as the smallest amount of analyte that can be quantified with acceptable accuracy. Precision is left to the judgment of the analyst, and the only requirement demanded is that any stated limit of detection be accompanied by precision and accuracy evaluation.

7.3.1. Mean Value

The mean or average value $(\overline{X})$ of a set of n values is determined by summing the individual and dividing by n measurements:

$$\overline{X} = \frac{\sum_{i=1}^{n} X_i}{n} \tag{7.13}$$

7.3.2. Standard Deviation

The standard deviation (σ) of a sample of n results or measurements is calculated by means of the following equation:

$$\sigma = \left[\frac{\sum_{i=1}^{n} X_i^2 - \frac{\left(\sum_{i=1}^{n} X_i\right)^2}{n}}{n-1} \right]^{1/2} \tag{7.14}$$

Variance is equal to σ^2.

7.3.3. Precision and Accuracy

Before beginning a discussion on the precision and accuracy aspects of gas chromatography, it is important that the two terms be defined.

Precision of data is best reported as relative standard deviation (RSD):

$$\text{RSD} = \frac{\text{standard deviation } (\sigma)}{\text{mean } (X)} \tag{7.15}$$

or often expressed as a percentage of the standard deviation

$$\%\text{RSD} = 100 \cdot \frac{\sigma}{X} \tag{7.16}$$

Relative standard deviation is also known as the coefficient of variation (CV). The precision with which a response ratio can be determined is found by repetitive gas chromatographic analysis, where the number of repetitions (n) are larger than or equal to 3 of any processed sample. Precision is increased by increasing the number of replicates, enabling an analyst to determine with greater statistical confidence that the true mean lies within certain limits about the experimental mean or to reduce the interval at a certain confidence level. Whenever precision is reported, it must be accompanied by an indication of the amount of analyte measured since precision is dependent on the amount of analyte. Precision represents a measure of the degree to which several runs are reproducible.

Confidence limits, or the confidence interval, is calculated as

$$\mu = \overline{X} \pm \frac{t \cdot s}{\sqrt{n}} \tag{7.17}$$

where μ is the true mean, X is the experimental value of the mean, and t is a value obtainable in Table 7.1 for different percentages of confidence and number of trials (n) or $\nu = n - 1$.

The value of t increases as the percentage confidence desired increases and decreases for more replicate analyses.

Accuracy of data refers to the difference between the experimentally determined mean value $\overline{X}$ and that value which is accepted as the true

TABLE 7.1
Values of t versus $\nu = n - 1$ for $\alpha = 0.05$ and 0.01

ν	$\alpha = 0.05$	t[a]	$\alpha = 0.01$
2	4.30		9.92
3	3.18		5.84
4	2.77		4.60
5	2.57		4.03
6	2.45		3.71
7	2.36		3.50
8	2.31		3.35
9	2.26		3.25
10	2.23		3.17

[a] Ref. 14.

value for the quantity measured. In cases where the true value is unknown, as is often the case in environmental analyses, accuracy is best evaluated by comparison of the GC results with those of another assay technique or with values from other laboratories using a chromatographic assay. Accuracy refers to the exactness or closeness of an answer to a known or actual value. Accuracy is usually expressed in terms of bias, error, or as percentage of recovery.

In order to reduce measurement error to tolerable limits and to provide a means of ensuring that the GC results have a high probability of being of acceptable quality, quality control and quality assurance programs are introduced.[14] Quality control is the program established to control errors, while quality assurance in environmental analysis represents a program which verifies that the gas chromatographic system operates within acceptable limits.

Quality assurance should be used by a laboratory to detect and correct problems and take measures to keep procedures reliable.[14] It requires properly organized laboratory facilities, validated analytical methodologies, and trained personnel. An important part of a quality assurance program requires the use of high-quality solvents, glassware, and standards. Further, it requires the employment of control and standard samples, the use of replicate samples, and comparison of replicate results with interlaboratory quality control samples. The accuracy of analytical data can be determined by participation in interlaboratory studies.

Let us consider the following 50 replicate results, expressed in ng/l, (ppt), obtained by analyzing a stable contaminant, such as 2,3,7,8-tetrachloro-dibenzo-p-dioxin in a sample.

Data (ng/l):

24.7	32.9	33.9	34.5	35.6
30.2	33.0	33.9	34.6	35.8
31.1	33.0	34.0	34.8	35.9
31.4	33.1	34.1	34.9	36.2
31.5	33.2	34.2	35.0	36.4
31.8	33.5	34.2	35.1	36.4
32.1	33.7	34.3	35.3	37.2
32.6	33.7	34.3	35.5	38.2
32.7	33.8	34.3	35.6	40.1
32.7	33.8	34.4	35.6	49.6

Mean X = 34.368 for 50 results

Mean of 48 results (omitting values of 24.7 and 49.6) = 34.2520

The two values 24.7 and 49.6 clearly indicate that the procedure is out of control; they are discarded. The value 40.1 is marginal and represents a more difficult decision. In this example it is left in provisionally. Justification for these statements is discussed in Sec. 7.3.4.

The estimate of the standard deviation, σ, is obtained:

$$\sigma^2 = \frac{\sum_{i=1}^{48} X_i^2 - n\overline{X}^2}{n-1} = \frac{56,470.35 - 48(34.2520)^2}{47} \tag{7.18}$$

$$\sigma = 1.8 \text{ ng/l}$$

It should be noted that the σ is expressed in absolute rather than relative terms (variance). If variability were proportional to concentration, then the relative standard deviation would be appropriate. However, if different ranges are used to determine the same component, an estimate of the standard deviation will be required for each range. One would not expect the variability which characterizes analyses in the range 1–100 ng to also apply to analyses in the range 0–100 μg.

7.3.4. Establishing Control Limits

There are two points in setting control limits. They should be close enough to signal when there is trouble with a system, and in contrast, they should be distant enough to discourage tinkering with a chromatographic system that is operating within its capabilities. The compromise solution which has been satisfactory in a great many applications is the use of 3σ control limits.

To decide whether the value 40.1 ng/l will remain in the data set for which an estimated standard deviation of 1.826 ng/l was calculated, the control limits are calculated.

$$34.2520 \pm (3 \times 1.8) = 39.7 \quad \text{and} \quad 28.7$$

Since 40.1 is larger than the upper control limit 39.7, there is sufficient evidence to discard this value also.

The estimate of the standard deviation is recalculated from the 47 data to give $\sigma = 1.6$ ng/l. The new sample mean is 34.128, resulting in new control limits of 29.3 and 39.0 ng/l which covers the 47 values remaining in the data set.[15]

7.4. PRIMARY STANDARDS IN AIR POLLUTION ANALYSIS

Permeation tubes have been developed as convenient sources of pollutants for preparing known low concentrations in air for calibration of instruments, development and testing of analytical methods, and toxicological studies. Liquidified gases or liquids such as hydrocarbons, mercaptans, and many others are sealed in Teflon tubes. The gases permeate out through the walls at a constant-mass permeation rate, generally a few milligrams a day.[16]

The quantitative relationships for permeation through a tube of unit length is derived by considering the mass flow through Teflon capillary tubing:

$$G = 37.68 \; P \cdot \sigma_0 r \; (-dp/dr) \cdot 10^6 \tag{7.19}$$

where G is the mass permeation rate in μg/min cm of tube length, P is the permeation constant for the gas through the Teflon in cm^3 (STP) cm/cm^2 sec (cm Hg), σ_0 is gas density g/cm^3 (STP) $= MW/22,414$, r is the radius of the Teflon tubing in cm, p is the gas pressure in mm Hg, and 10^6 is a constant for converting units of mass, pressure, and time.

Under steady-state conditions, the mass permeation rate is constant through the wall and after separation of variables p and r and subsequent integration

$$G = 730P \cdot M \cdot p_1 / \log(d_2/d_1) \qquad (7.20)$$

where M is the molecular weight of the gas, d_2 is the outside diameter of the permeation tube, d_1 is the inside diameter of the tube and p_1 is the gas pressure inside the tube in mm Hg.

The permeation rate depends upon the ratio of outside-to-inside diameter of the tube. The absolute diameter does not affect the permeation rate. On the other hand the volume of the entraped liquidified gas and hence the lifetime of the permeation tube is proportional to the square of the inside diameter.[17]

The permeation rates have high-temperature coefficients, and also an appreciable activation energy, which is higher than the heat of vaporization of the liquidified gas. The gravimetric permeation rates at different temperatures can be calculated as

$$\log\left(\frac{G_2}{G_1}\right) = \frac{E}{2.303R}\left(\frac{1}{T_1} - \frac{1}{T_2}\right) \qquad (7.21)$$

TABLE 7.2
Permeation Rates of Some Common Industrial Chemicals at 70 °C

Contaminant	Rate (ng/min cm)	K
Acetone	304	0.422
Benzene	207	0.313
Carbon tetrachloride	265	0.159
Chloroform	715	0.205
Cyclohexane	20	0.291
1,2-dichloroethane	331	0.247
Ethylbenzene	35	0.231
n-Hexane	162	0.284
Methylene chloride	1690	0.288
Methyl ethyl ketone	140	0.340
Methanol	245	0.764
1,1,1-trichloroethane	113	0.183
Trichloroethylene	1060	0.186
Toluene	120	0.266
Vinyl chloride	460	0.391
Vinyl acetate	450	0.284
o-Xylene	37	0.231

where G_1 and G_2 are gravimetric permeation rates, T_1 and T_2 are corresponding temperatures in °K, E is the activation energy of the permeation process in cal/gmol, and R is the gas constant in 1.987 cal/gmol °K. Table 7.2 lists permeation rates of some common industrial contaminants.

A variety of liquefiable gases can be sealed in permeation tubes to provide constant sources of small quantities of analytes. The tubes can be calibrated gravimetrically or volumetrically. They may be used conveniently under themostated conditions in the flow dilution system to provide accurately known low concentrations of analytes.[18]

REFERENCES

1. J. Novak, *Quantitative Analysis by Gas Chromatography*, Dekker, New York, 1975.
2. K. K. Carroll, *Nature* **191**, 337 (1961).
3. J. C. Bartlet and J. L. Iverson, *J. Assoc. Offic. Anal. Chem.* **49**, 21–27 (1966).
4. A. E. Brandt and W. E. M. Lands, *Lipids* **3**, 179 (1967).
5. R. W. Spillman and H. V. Malmstadt, *Anal. Chem.* **48**, 303–311 (1976).
6. A. Shatkay, *Anal. Chem.* **50**, 1423–1424 (1978).
7. S. Dal Nogare and R. S. Juvet, *Gas-Liquid Chromatography, Theory and Practice*, Interscience, New York, Chapter VII, 1962.
8. A. Shatkay, *J. Chromatogr.* **198**, 7–22 (1980).
9. J. Drozd and J. Novak, *J. Chromatogr.* **165**, 141–165 (1979).
10. W. J. Khazal, J. Vejrosta, and J. Novak, *J. Chromatogr.* **157**, 125–131 (1978).
11. G. Widmark, Proceedings of the International Symposium on Identification and Measurement of Environmental Pollutants, Ottawa, Ontario, June, 1971, p. 396.
12. J. L. Clark, *The Quality Control Handbook for Pilot Watershed Studies*, (PLUBAG), International Joint Commission, Windsor, Ontario.
13. J. K. Taylor, *Anal. Chem.* **53**(14), 1588A–1596A (1981).
14. K. Eckschlager, *Graphical Methods in Analytical Chemistry*, Státní Nakladatelství Technické Literatury, Prague, 1966, p. 101.
15. *ASTM Manual on Presentation of Data and Control Chart Analysis*. ASTM Special Technical Publication, 15D, 1976.
16. A. E. O'Keeffe and G. C. Ortman, *Anal. Chem.* **38**, 760–763 (1966).
17. B. E. Saltzman, *Proceedings of the International Symposium on Identification of Environmental Pollutants*, (B. Westley, ed.), Ottawa, 1971, p. 64.
18. F. J. Debrecht in *Modern Practice of Gas Chromatography* (R. L. Grob, ed.), Wiley, New York, 1977.

APPLICATION IN ENVIRONMENTAL ANALYSIS

8.1. INTRODUCTION

The application of WCOT column gas chromatography for use in the environmental laboratory is relatively new. This is partly due to the shortage of qualified personnel to standardize and perfect WCOT column gas chromatography for routine environmental laboratory procedures. This situation occurred even though highly selective and specific methodology of WCOT column gas chromatography has been available for many environmental applications for some years now.

The practice of high-resolution gas chromatography requires the knowledge of the fundamentals discussed in the previous chapters. At present, few environmental laboratories use high-resolution gas chromatographic methods because environmental analytical chemists have come to rely on thoroughly tested but obsolete procedures previously employed by pesticide chemists. Of course, the principles of such methods can be used for the determination and characterization of samples, but the complexity of environmental samples require much more powerful separation efficiency than packed column gas chromatography is able to provide. Figure 8.1 illustrates this.

Less than a decade ago it was difficult to find more than a few environmental applications of high-resolution gas chromatography. Today, the widespread use of this technique has only just begun. Many environmental analytical chemists are already aware of its considerable potential. Virtually every branch of chemistry makes extensive use of high-resolution WCOT column gas chromatography. The utility of high-resolution gas chromatography coupled with mass spectrometry (GC/MS) has been realized by those working in environmental laboratories, and it is now a standard technique of analysis. The application of any method of analysis for one or more components is successful only insofar as the analytical chemist can cope with the matrix in which the compound is dispersed. This is particularly true in environmental analyses where the matrices might either be biological tissue, water, air and sediment, or biota and fauna.

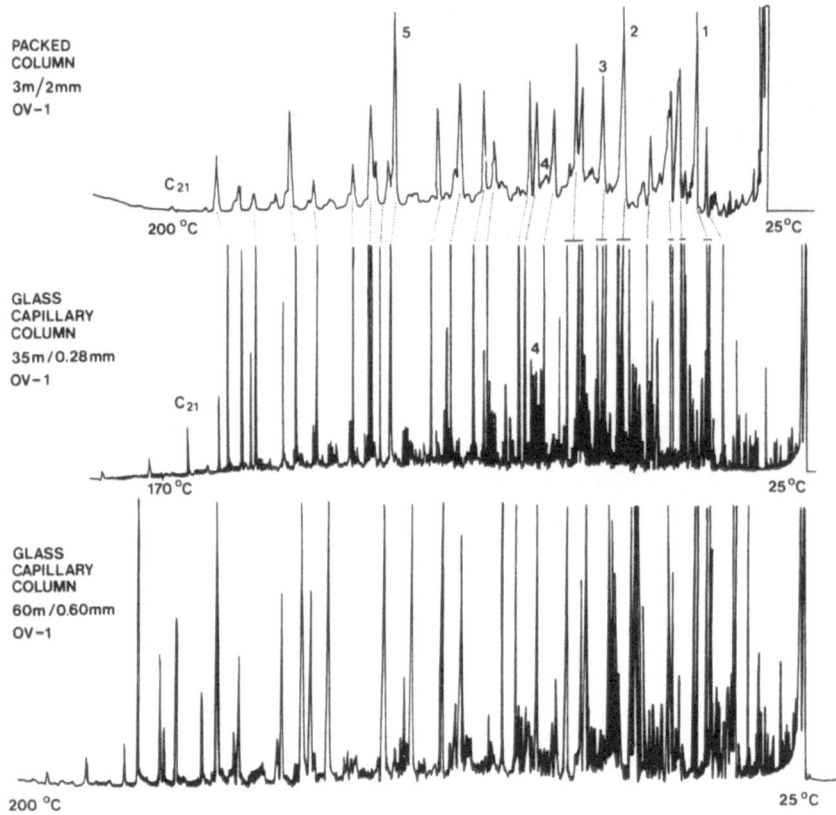

FIGURE 8.1. Comparison of typical chromatograms from packed and WCOT columns. The injected environmental sample and stationary phase OV-1 are the same for all columns. (Courtesy of Prof. K. Grob, EAWAG, Dübendorf, Switzerland.)

The problems facing the analyst in developing a high-resolution gas chromatographic method are not unique. With a few exceptions, the logical basis for method development in environment analyses by gas chromatography is similar to many other fields. The following applications have been chosen because they show a specific point of the technique, they are of general interest to the environment analysts, or because of the impact that the high-resolution WCOT column gas chromatography or GC/MS have on environmental analytical chemistry.

8.2. ANALYSIS OF ORGANIC PRIORITY POLLUTANTS IN WATER

There are four categories of organic priority pollutants: volatiles, base–neutral extractables, acid extractables, and pesticides. Each category

TABLE 8.1
Characteristic Ions of Volatile Purgable Organic Pollutants

Compound	Ion at m/z (relative intensity)	Ion for SIM (m/z)
Chloromethane	50(100); 52(33)	50
Dichlorodifluoromethane	85(100); 87(33); 101(13); 103(10)	101
Bromomethane	94(100); 96(94)	94
Vinyl chloride	62(100); 64(33)	62
Chloroethane	64(100); 66(33)	64
Methylene chloride	49(100); 51(33); 84(86)	84
Trichlorofluoromethane	101(100); 103(66)	101
1,1-dichloroethylene	61(100); 96(80); 98(53)	96
Bromochloromethane	49(100); 130(88); 128(70)	128
1,1-dichloroethane	63(100); 65(33); 83(13)	63
trans-1,2-dichloroethylene	61(100); 96(90); 98(57)	96
Chloroform	83(100); 85(66)	83
1,2-dichloroethane	62(100); 64(33); 98(23)	96
1,1,1-trichloroethane	97(100); 99(66); 117(17)	97
Carbon tetrachloride	117(100); 119(96); 121(30)	117
Bromodichloromethane	83(100); 85(66); 127(13)	127
1,2-dichloropropoane	63(100); 65(33); 112(4)	112
trans-1,3-dichloropropene	75(100); 77(33)	75
Trichloroethylene	95(100); 97(66); 130(90); 132(85)	130
Dibromochloromethane	129(100); 127(78); 208(13); 206(18)	127
cis-1,3-dichloropropene	75(100); 77(33)	75
1,1,2-trichloroethane	97(100); 83(95); 99(63); 85(60)	97
Benzene	78(100)	78
2-bromo-1-chloroethane	63(100); 65(33); 142(14); 144(18)	142
Bromoform	173(100); 171(50); 175(18); 254(11)	173
1,1,2,2-tetrachloroethene	129(00); 131(59); 166(100); 164(78)	164
1,1,2,2-tetrachloroethane	83(100); 85(66); 133(7); 168(6)	168
1,4-dichlorobutane	55(100); 90(21); 92(7)	90
Toluene	91(100); 92(78)	92
Chlorobenzene	112(100); 114(33)	112
Ethylbenzene	91(100); 106(33)	106
Acrolein	27(100); 26(49); 56(83); 55(64)	56
Acrylonitrile	26(100); 53(99); 52(75); 51(32)	53

requires a different analytical procedure.[1] Priority pollutant substances (toxic pollutants or consent decree pollutants) were selected on the basis of their known occurrence in drinking water, industrial effluents, fish, sediment, sludge and sewage, their persistence in the aquatic food web, their suspected mutagenic, carcinogenic, or teratogenic activity, and ability to bioaccumulate in our environment.

Normally, organic pollutants are presented in concentrations of $\mu g/l$ (or $\mu g/kg$) or lower in environmental samples, and they must be preconcentrated prior to chromatographic analysis. The volatile organics are

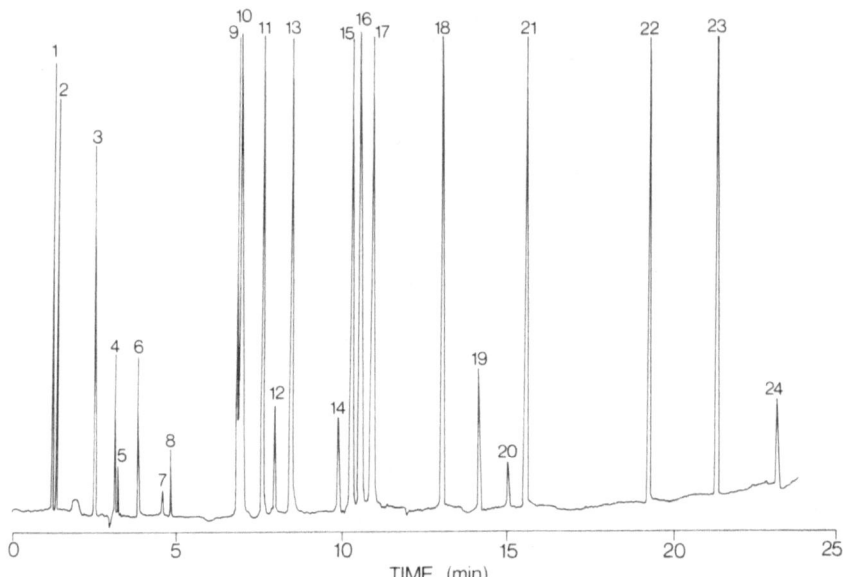

FIGURE 8.2. Chromatogram of volatile priority pollutants using ECD peaks: (1) oxygen; (2) dichlorodifluoromethane; (3) trichlorofluoromethane; (4) vinylidene chloride; (5) carbon disulfide; (6) methylene chloride; (7) 1,2-dichloroethane; (8) 1,1-dichloroethane; (9) bromochloromethane; (10) chloroform; (11) 1,1,1-trichloroethane; (12) 1,2-dichloroethylene; (13) carbon tetrachloride; (14) 1,2-dichloropropane; (15) trichloroethylene; (16) dibromomethane; (17) dichlorobromomethane; (18) trichlorobromomethane; (19) 1,1,2-trichloroethane; (20) dibromochloromethane; (21) tetrachloroethylene; (22) bromoform; (23) s-tetrachloroethane. (Courtesy of M. Comba and Dr. K.L.E. Kaiser.)

purged from water samples and are trapped on solid adsorbents such as Tenax-GC (60–80 mesh). The recommended purge–trap methodology has been described.[2,3] After the organic contaminants are trapped, they are rapidly desorbed using heat from the trap directly onto the chromatographic column.

The characteristic masses or mass ranges recommended for use by the US EPA are listed in Table 8.1. They are used for qualitative and quantitative determination of volatile priority pollutants.[4]

Acrolein and acrylonitrile are not efficiently recovered by the purge–trap method and must be determined by direct aqueous injection employing GC/MS. In general, direct aqueous injection is recommended for all compounds that exceed 1 mg/l. Figure 8.2 shows the volatile priority pollutants separated on a 25 m × 0.2-mm OV-101 meter glass WCOT column using temperature programming from −20 to 80 °C at 4 °C/min.

The procedure used for high-resolution GC/MS operation employing stripping and trapping may be described in the time and temperature scheme shown in Table 8.2.Time counting starts when the needle of the desorption device penetrates through the septum and the back flushing of the trap with helium is initiated. The computer acquisition could be started even earlier than indicated in Table 8.2. The ion source is turned on immediately after the desorption unit is withdrawn from the gas chromatograph.

It is recommended to perform three consecutive 30-min purges with helium at three different temperatures.[5] Identification of the volatile organics is confirmed by running standard compounds on the same column and under the same analytical conditions. The relative retention times and mass spectra of these standards are compared with those compounds detected in the unknown sample. It is also noted whether the mass spectrum of the standard and that of the recorded compound in the sample agrees with a mass spectrum published in the literature.[6]

Volatile halogenated hydrocarbons can be quantitatively determined using a chromatographic system consisting of the glass capillary WCOT column coated with a 1-μm film thickness of SE-52, 30 m in length, and the electron capture detector.[7] Short and very long WCOT columns coated with SE-54, OV-17, and SP-2100 have been employed for the analysis of volatile organic priority pollutants.[8] Up to 16 extracts per hour can be determined with reproducibility better than 2%. Simultaneous analysis of all five organic priority pollutant fractions containing acrylonitrile, acrolein, base–neutral extractables, and pesticides are chromatographed in a single run by injecting the extractable organic pollutants, then applying a cryogenic focusing to the front of the WCOT column, and desorbing the volatile components. The WCOT column used is a fused-silica 30 m × 0.25-mm i.d. coated with SE-54. The column is directly interfaced to the ion source of the mass spectrometer.[9]

TABLE 8.2
Operational Parameters for Stripping–Trapping of Volatile Pollutants

Operation	Time (min)	GC temperature (°C)
Purge and trap	4	20
Raise oven temperature to 120 °C	3	20–120
Start acquisition	3	120
Hold at 120 °C	6	120
Start temperature programming		
Terminate run		

TABLE 8.3
GC/MS Data for Analysis of Base-Neutral Extractables

Compound name	RRT	DL (ng)	Ions-m/z (relative intensity)
1,3-dichlorobenzene		40	146(100); 148(64); 113(12)
1,4-dichlorobenzene		40	146(100); 148(64); 113(11)
Hexachloroethane		40	117(100); 199(61); 201(99)
1,2-dichlorobenzene		40	146(100); 148(64); 113(11)
bis[2-chloroisopropyl] ether		40	45(100); 77(19); 79(12)
Hexachlorobutadiene		40	225(100); 223(63); 227(65)
1,2,4-trichlorobenzene		40	74(100); 109(80); 145(52)
Naphthalene	0.352	40	128(100); 127(10); 129(11)
bis[2-chloroethyl] ether		40	93(100); 63(99); 95(31)
Hexachlorocyclopentadiene		40	237(100); 235(63); 272(12)
Nitrobenzene		40	77(100); 123(50); 65(15)
bis[2-chloroethoxy] methane		40	93(100); 95(32); 123(21)
2-chloronaphthalene		40	162(100); 164(32); 127(31)
Acenaphthylene	0.720	40	152(100); 153(16); 151(17)
Acenpahthene	0.768	40	154(100); 153(95); 152(53)
Isophorone		40	82(100); 95(14); 138(18)
Fluorene	0.926	40	166(100); 165(82); 167(13)
2,6-dinitrotoluene		40	165(100); 63(72); 121(23)
1,2-diphenylhydrazine		40	77(100); 93(58); 105(28)
2,4-dinitrotoluene		40	165(100); 63(72); 121(23)
Hexachlorobenzene		40	284(100); 142(30); 249(24)
4-bromophenylphenyl ether		40	248(100); 250(96); 141(45)
Phenanthrene	1.189	40	178(100); 179(16); 176(15)
Anthracene	1.203	40	178(100); 179(16); 176(11)
Dimethylphthalate		40	163(100); 164(10); 194(11)
Diethylphthalate		40	149(100); 178(25); 150(10)
Fluoranthene	1.546	40	202(100); 101(23); 100(14)
Pyrene	1.603	40	202(100); 101(26); 100(17)
Di-n-butylphthalate	1.453	40	149(100); 150(27); 104(10)
Benzidine		40	184(100); 92(24); 185(13)
Butylbenzylphthalate		40	149(100); 91(50)
Chrysene		40	228(100); 229(19); 226(23)
bis[2-ethylhexyl]phthalate		40	149(100); 167(31); 279(26)
Benzo[a]anthracene		40	228(100); 229(19); 226(19)
Benzo[b]fluoranthene		40	252(100); 253(23); 125(15)
Benzo[k]fluoranthene		40	252(100); 253(23); 125(16)
Benzo[a]pyrene		40	252(100); 253(23); 125(21)
Indeno[1,2,3-c,d]pyrene		100	276(100); 138(28); 277(27)
Dibenzo[a,h]anthracene		100	278(100); 139(24); 279(24)
Benzo[ghi]perylene		100	276(100); 138(37); 277(25)
N-nitrosodimethylamine			42(100); 74(88); 44(21)
N-nitrosodi-n-propylamine			130(100); 42(64); 101(12)
4-chloro-phenylphenyl ether			204(100); 206(34); 141(29)
Endrin aldehyde			67(100); 250(40); 345(32)
3,3'-dichlorobenzidine			252(100); 254(66); 126(16)
bis(chloromethyl) ether			45(100); 49(14); 51(05)
d_{10}-anthracene		40	188(100); 94(19); 80(19)
2,3,7,8-tetrachlorodibenzo-p-dioxin			322(100); 320(90); 257(22)

8.2.1. Nonvolatile Pollutants

This group covers those priority pollutants associated with the consent decree that are solvent extractable and amenable to gas chromatography. These pollutants are listed in Tables 8.3–8.5.

A GC/MS method is most usually employed for qualitative and quantitative determination and confirmation of these contaminants in environmental samples. Pesticides are also qualitatively confirmed by GC/MS. They can be determined quantitatively by electron capture detection using WCOT column gas chromatography.

8.2.2. Base–Neutral Extraction

Water or a composite sample including a representative portion of the suspended solids can be extracted in a 4-l separatory funnel or a continuous extractor. The pH of the sample is adjusted to 12 with 6 N sodium hydroxide using pH multirange paper for the measurement. Samples are serially extracted with 250-, 100-, and 100-ml portions of methylene chloride. Each portion of the sample is shaken for at least 2 min with methylene chloride. Afterward, it is dried and filtered through a short column of sodium sulfate and concentrated by Kuderna–Danish

TABLE 8.4
GC/MS Data for Analysis of Acid-Extractable Priority Pollutants

Compound name	LD (ng)	Ions-m/z (relative intensity)	CI ions
2-chlorophenol	100	128(100); 64(54); 130(31)	129; 131; 157
Phenol	100	94(100); 65(17); 66(19)	95; 123; 135
2,4-dichlorophenol	100	162(100); 164(58); 98(61)	163; 165; 167
2-nitrophenol	100	139(100); 65(35); 109(08)	140; 168; 122
p-chloro-m-cresol	100	142(100); 107(80); 144(32)	143; 171; 183
2,4,6-trichlorophenol	100	196(100); 198(92); 200(26)	197; 199; 201
2,4-dimethylphenol	100	122(100); 107(90); 121(55)	123; 151; 163
2,4-dinitrophenol	100	184(100); 63(59); 154(53)	185; 213; 225
4,6-dinitro-o-cresol	100	198(100); 182(35); 77(28)	199; 227; 239
4-nitrophenol	100	65(100); 139(45); 109(72)	140; 168; 122
Pentachlorophenol	100	266(100); 264(62); 268(63)	267; 265; 269
d_{10}-anthracene	40	188(100); 94(10); 80(18)	189; 217

distillation. The extract is preconcentrated to 5–10 ml using a three-ball macro-Snyder column and a 10-ml calibrated tube.

The concentrate is further preconcentrated to 1 ml, and 10 μl of 2 μg/μl of d_{10}-anthracene is added as the internal standard. The base–neutral extractables may be separated in the GC/MS system using either a 35 m $\times$ 0.25-mm i.d. SP-2250 or 35 m $\times$ 0.25-mm SP-2100 WCOT column with temperature programming from 80 to 240 °C at 4 °C/min.

A surface water extract is shown in Fig. 8.3. The confirmation is based on retention data and selected ion monitoring.[10] Confidence in the data can be improved by comparing data from two WCOT columns.[11]

8.2.3. Acid-Extractable Priority Pollutants

The pH of the base–neutral extracted water is adjusted with 6 N HCl to pH 2 or less. The sample is extracted with a series of 200-, 100-, and 100-ml portions of methylene chloride in the same manner as for the base–neutral fraction including the addition of the internal standard. The recovery of 85% of the added solvent constitutes a working definition of having broken any emulsion formed.

A total ion chromatogram for GC/MS analysis of acid-extractable priority pollutants using a 30-m 0.25-mm i.d. SE-54 WCOT column is shown in Fig. 8.4. The splitless injection technique was used and helium employed as the carrier gas with linear velocity of 20 cm/sec. Temper-

TABLE 8.5
Pesticides Extractables

Organochlorine compound	m/z	Organochlorine compound	m/z
α-hexachlorocyclohexane	181	o,p-DDD	235
Hexachlorobenzene	284	Endrin	81
β-Hexachlorocyclohexane	181	p,p'-DDD	235
γ-hexachlorocyclohexane	181	p,p'-DDT	235
Heptachlor	100	Endosulfane sulfate	272
Aldrin	66	PCBs	
γ-chlordane	393	Toxaphene	159
α-chlordane	373	Tetrachlorinated dibenzo-p-dioxins	322
Dieldrin	79	Tetrachlorinated dibenzo-furans	306
p,p'-DDE	246		

ature programming was initiated using a 2-min hold at the initial temperature of 35 °C and programmed to 260 °C at 10 °C/min.[12]

8.2.4. Pesticides

These compounds are analyzed by means of electron capture detection gas chromatography (GC–ECD). Usually a 1-l sample, well homogenized with a suspended sediment, is transferred to a 2-l separatory funnel, and the cylinder is rinsed with the first portion of the extracting solvent. The solvent amounts remain unchanged. The methylene chloride solvent must be evaporated to a small volume and exchanged into n-hexane for cleanup or GC–ECD analysis. The evaportion to 5–8 ml can be performed using a rotary evaporator. Afterward, 20 ml of n-hexane is added along with a fresh portion of boiling stones, and the sample is evaporated to the desired volume, usually 1–5 ml or less.

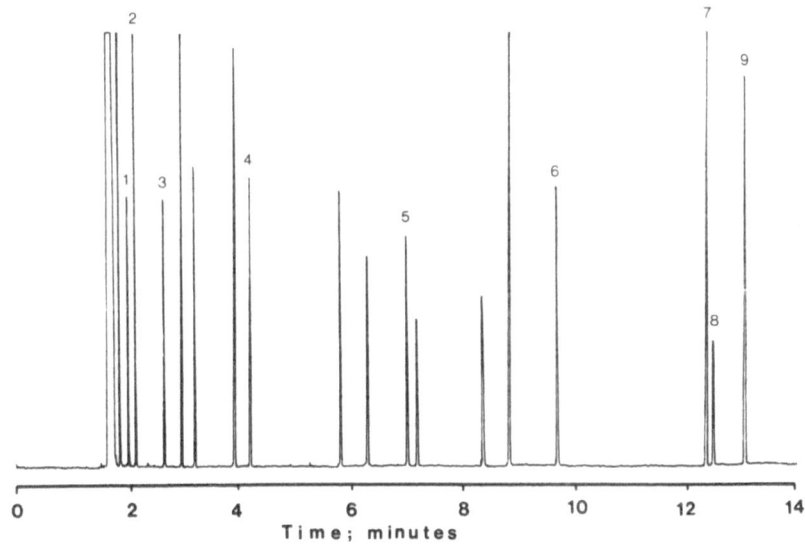

FIGURE 8.3. The chromatogram of selected pollutants: (1) methylene chloride; (2) trans-1,2-dichloroethylene; (3) chloroform; (4) bromodichloromethane; (5) chlorodibromomethane; (6) bromoform; (7) 1,3-dichlorobenzene; (8) 1,4-dichlorobenzene; (9) 1,2-dichlorobenzene on 60 m × 0.2-mm i.d. SE-54 WCOT column. Temperature programmed from 35 to 120 °C at 10 °C/min using hydrogen as a carrier gas.

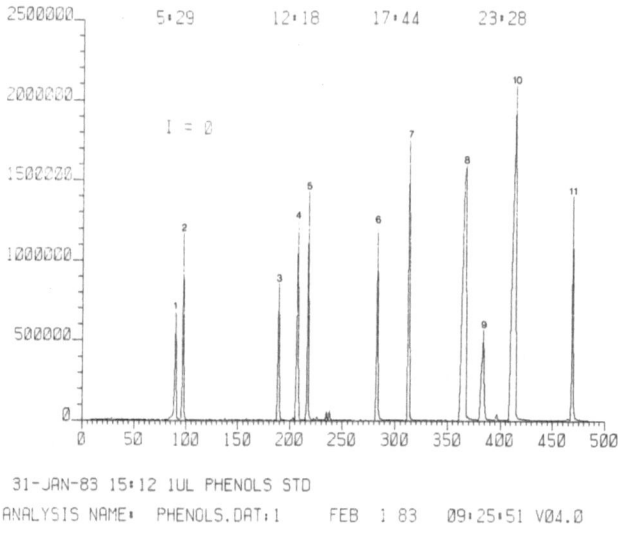

FIGURE 8.4. Acid-extractable priority pollutant standard mixture recorded by total ion monitoring: (1) phenol; (2) 2-chlorophenol; (3) 2,4-dichlorophenol; (4) 2-nitrophenol; (5) 4-chloro-3-cresol; (6) 2,4,6-trichlorophenol; (7) 2,4-dimethylphenol; (8) 2,4-dinitrophenol; (9) 4,6-dinitro-2-cresol; (10) 4-nitrophenol; (11) pentachlorophenol.

Eighteen pesticides and polychlorinated biphenyls (PCB) are confirmed by GC/MS. Chlordane, toxaphene, and the PCB compounds represent multicomponent mixtures and are usually characterized using retention windows rather than specific retention times. The last two mixtures require special treatment, and methods for them are not finalized yet. Their separation and confirmation will be described separately.

Figure 8.5 depicts an analysis of a series of pesticide standards using the negative ion chemical ionization mode and methane as reagent gas. The column used is a 25 m × 0.3-mm i.d. fused-silica WCOT column coated with OV-101 methyl silicone oil. A quadrupole mass spectrometer is scanned from 100 to 450 amu with a scan cycle time of 0.9 sec.[13] The peak shape and resolution for all pesticides is excellent. However, sensitivity varies and responses are different from compound to compound. The most sensitive responses were obtained for hexachlorobenzene and chlordanes.

The analytical scheme is more complex when a large number of extractable organics in samples such as sludges and municipal wastewater must be analyzed. Special extraction and cleanup procedures are required before WCOT GC/MS analysis can be employed.[13] The interferences

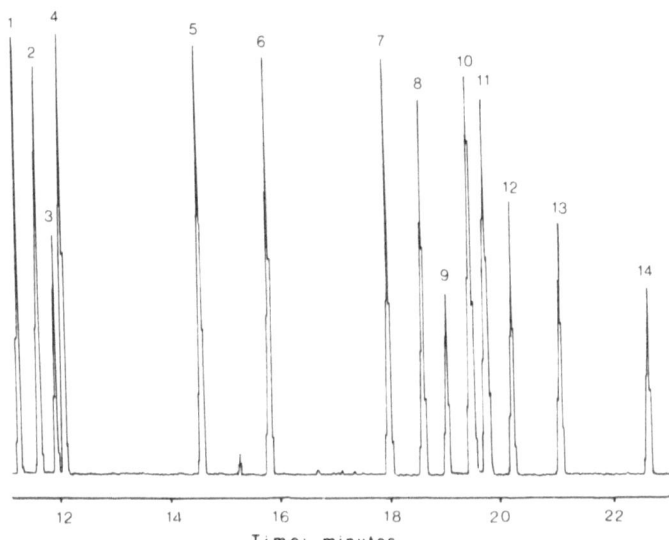

FIGURE 8.5. Chromatogram of pesticide standards (30 ng each) on 25 m × 0.3-mm i.d. OV-101 fused-silica WCOT column. (1) α-Hexachlorocyclohexane; (2) hexachlorobenzene; (3) β-hexachlorocyclohexane; (4) γ-hexachlorocyclohexane; (5) heptachlor; (6) aldrin; (7) β-chlordane; (8) α-chlordane; (9) dieldrin; (10) *p,p'*-DDE; (11) *o,p'*-DDD; (12) endrin; (13) *p,p'*-DDD; (14) *p,p'*-DDT.

extracted from municipal wastewater and sludge samples contain lipids, fatty acids, and saturated hydrocarbons. They must be removed using the conventional approaches of acid–base separation, molecular size separation by means of gel permeation chromatography, and column chromatography using polarity separation principles.

Acid–base separation is the fundamental separation principle. In this methodology base–neutral extraction, followed by extraction with acid, divides interferences between base and acid extracts. It separates the base–neutral extractables from the acid extractables, and thus reduces the degree of interference in each fraction.

Molecular size separation is effective in removing the lipids, high molecular weight fatty acids, and hydrocarbons from extracts. It is important to remove these materials from the extract in order to extend WCOT column life.

Column chromatography with silica gel, Florisil, and cesium silicate is employed to separate saturated hydrocarbons from the aromatics and polar compounds such as phenols from neutral interferences. A method has been developed for the analysis of the extractable priority organic pollutants in complex samples by DeWalle and Chian.[14]

GC/MS analysis of composite standards and extracts show that WCOT column separation may provide quantitative results even for simultaneous analysis of all five organic priority pollutant fractions. The chromatogram obtained by this technique is shown in Fig. 8.6.

8.3. DETERMINATION OF GROUPS OF COMPOUNDS

The efficiency of WCOT columns, besides being essential for the analysis of the complex mixtures obtained from environmental and biological samples, also makes them ideal for the analysis of specific compounds or groups of compounds such as aliphatic and aromatic hydrocarbons, phenols, chlorinated dibenzo-*p*-dioxins, polychlorinated

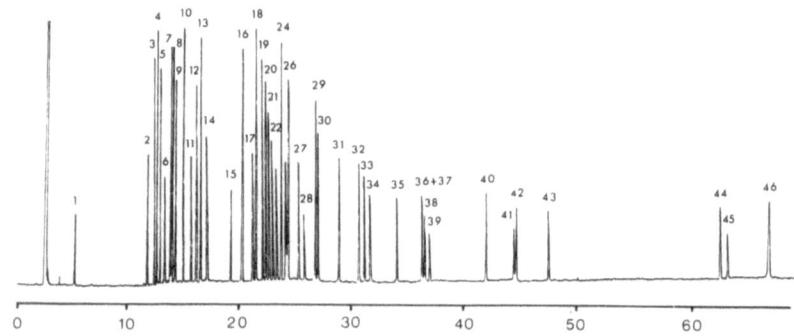

FIGURE 8.6. Chromatogram of base–neutral priority pollutants on 30 m × 0.2-mm SE-54 bonded FS-WCOT column. Temperature programmed from 50 to 270 °C at 8 °C/min; helium as a carrier gas $u_0 = 20$ cm/sec. Peak identification: (1) *N*-dimethylnitrosoamine; (2) bis(2-chloroethyl) ether; (3) 1,3-dichlorobenzene; (4) 1,4-dichlorobenzene; (5) 1,2-dichlorobenzene; (6) bis(2-chloroisopropyl) ether; (7) hexachloroethane; (8) *N*-nitroso-di-*n*-propylamine; (9) nitrobenzene; (10) isophonone; (11) bis(2-chloroisopropyl) ether; (12) 1,2,4-trichlorobenzene; (13) naphthalene; (14) hexachlorobutadiene; (15) hexachlorocyclopentadiene; (16) 2-chloronaphthalene; (17) acenaphthylene; (18) dimethylphthalate; (19) 2,6-dinitrotoluene; (20) acenaphthene; (21) 2,4-dinitrotoluene; (22) 4-chlorophenylphenyl ether; (23) fluorene; (24) diethylphthalate; (25) *N*-nitrosodiphenylamine; (26) 1,2-diphenylhydrazine; (27) 4-bromophenylphenyl ether; (28) hexachlorobenzene; (29) phenanthrene; (30) anthracene; (31) di-*n*-butylphthalate; (32) fluoranthene; (33) benzidine; (34) pyrene; (35) butylbenzyl phthalate; (36) 3,3'-dichlorobenzidine; (37) chrysene; (38) benzo[*a*]anthracene; (39) bis(2-ethyl hexyl) phthalate; (40) di-*n*-octylphthalate; (41) benzo[*b*] fluoranthene; (42) benzo[*k*]fluoranthene; (43) benzo[*a*] fluoranthene; (44) indeno[1,2,3-*c,d*]pyrene; (45) dibenz[*a,h*]anthracene; (46) benzo[*ghi*]perylene.

biphenyls, dibenzofurans, polychlorinated camphenes, and many other similar mixtures. The efficiency of WCOT columns facilitates the separation of structurally similar congeners of such groups and, in addition, permits separation of compounds from sample interferences.

Aliphatic and aromatic hydrocarbons are probably the most studied group of organic compounds because of the continuing threat of pollution by them. There have been many reviews devoted to analyses of hydrocarbons. Because of their nonpolar nature, separation methods for aliphatic hydrocarbons are simple.[15,16] Hydrocarbons may be present in water in true solution, in suspension as particulate or as droplets. In all of these forms they may be separated from water by extraction with organic solvents. There is no single well-accepted extraction technique for hydrocarbons. To some extent, the choice of solvent may depend on the analytical technique to be employed for subsequent measurement. The most commonly employed solvents are hexane and carbon tetrachloride.[17,18]

Concern over contamination of inland waters and oceans by hydrocarbons has led to the application of WCOT column techniques to the qualitative and quantitative determination of aliphatic and aromatic hydrocarbons in aquatic environment. The majority of the studies have been carried out on a qualitative basis and have been directed toward the identification of hydrocarbon pollutants by the comparison of chromatographic profiles of unidentified hydrocarbons either visually[19] or by fingerprinting.[20,21]

WCOT gas chromatography has been used mainly for analysis of hydrocarbon pollution in relation to inland waters and marine pollution incidents.[22,23] These columns have also been used for the analyses of hydrocarbons in sediments.[24,25]

The analysis of hydrocarbons is complicated both by the different types of original crude oil and by the alterations to the profiles caused by weathering. These changes are caused by adsorption, biological action, photodegradation, and evaporation. The chromatographic profiles of the water extracts may be compared with those from crude oil samples which have been naturally or artificially weathered.[23,26] However, oils detected in groundwater are usually different from the original matrix due to biodegradation and adsorption of hydrocarbons.[27,28] A wide variety of different types of WCOT columns has been applied, although the majority of studies have been carried out using glass columns.[29–33]

Analyses are usually carried out using nonpolar stationary phases such as OV-1, SE-52, OV-101, Apiezon L, and recently introduced bonded phases such as DB-1 and DB-5 which are suitable for high column temperatures.

The determination, both qualitative and quantitative, of the aliphatic and aromatic hydrocarbons by WCOT column gas chromatography has seemed to be an obvious development ever since the invention of the flame ionization detector (FID). The use of sulfur-specific flame photometric and nitrogen-sensitive FID has been employed in conjunction with FID to provide more information. The sulfur-specific detector is especially useful for the identification of weathered samples because the sulfur-containing compounds are biodegraded less rapidly then the hydrocarbons.

The sensitivity of FID in many instances is not great enough to detect the hydrocarbons in unpolluted waters and for this reason GC/MS, especially when employing chemical ionization mode of operation, provides the increased sensitivity needed. As an example, a gas chromatogram of oil spill samples is presented in Fig. 8.7. The loss of volatile components can be evaluated by inspecting the gas chromatogram. Light components of iso- and n-alkanes lost by evaporation from the AMOCO oil spill in Brettany represent about 12% of original oil, whereas light aromatic component loss is estimated to 20%. These data indicate that 60,000–70,000 tons of aliphatic hydrocarbons and up to 40,000 tons of volatile organic components, mainly benzene and higher methylated isomers, contaminated water, sediment, biota, and atmosphere.[21]

Quantitation of hydrocarbons by functional group type such as gasolines, naphthas, and refinery process streams are traditionally evaluated according to the amounts of hydrocarbons of each functional group giving the amounts of n-alkanes, n-alkenes, isoalkanes, naphthenes, and aromatics.

The sample is separated into its individual components by WCOT column gas chromatography, each component is quantitated, and then the sum of the amounts of individual components is added. A very detailed analysis is provided by this technique, and the response factors are stable and predictable.[34]

A number of review articles covering various aspects of hydrocarbon analyses have appeared. Desty covered general aspects,[35] and a review on the analysis of commercial C_4 fractions,[36] petroleum hydrocarbons,[37] modern methods of analysis,[38] and pollution control[39–41] all have been reported. A gas chromatographic technique has been developed for the determination of individual paraffin, olefin, naphthene, and aromatic constituents.[42] The method has been used to show that methanol-derived gasolines are much higher in olefins and naphthenes. Olefins in motor gasolines were determined using WCOT column gas chromatography with a hydrogenation precolumn containing aged Pd catalyst and an

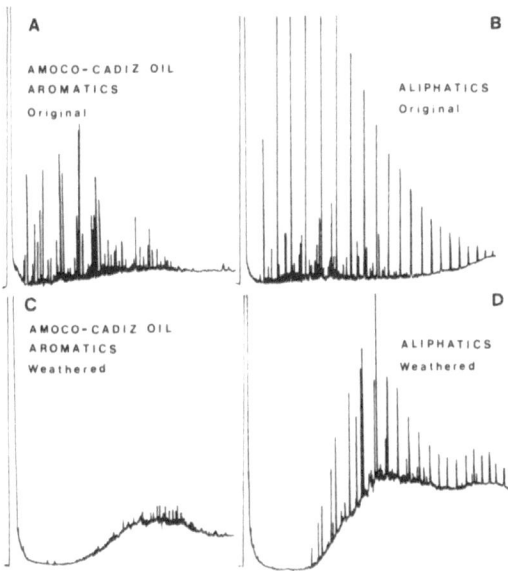

FIGURE 8.7. Chromatograms of aliphatic and aromatic hydrocarbons of (A, B) original and (C, D) weathered oil. Column: 50 m × 0.3-mm i.d. coated with SE-52. Splitless mode. Temperature programmed from 40 to 80 °C ballistically and then at 4 °C/min to 280 °C. Hydrogen carrier gas.

absorption precolumn containing H_2SO_4, H_3PO_4 coated Chromosorb, and optionally $HgSO_4$.[43] The WCOT column coated with polyphenyl ether for C_6–C_{14} n-alkanes separation was reported.[44] The same authors also published a series of studies on n-alkynes and monosubstituted cycolalkanes.[45–48] The best resolution of positional isomers has been obtained on squalane. Black described a direct analysis method for nonreactive hydrocarbons in air,[49] and a similar rapid method of analysis for C_2–C_5 alkanes has been described.[50] Quantitation by GC/MS using deuterated analogs has been reported for environmental alkanes[51] and for water-soluble gasoline fractions.[52]

The determination of retention indices of a series of alkylbenzenes on several liquid phases was reported.[15] Some problems encountered with the precision of the retention indices of alkylbenzenes in relation to the method of retention time measurement, nonideality of the carrier gas, column temperature, and aging of the column were discussed by Sojak.[54,55]

The closed-loop stripping method applied to an accidental diesel oil spill site in Switzerland has been described by Grob.[56] Two WCOT col-

were used in tandem with mass spectrometry to identify diesel oil components in groundwater, containing both n-alkanes and alkylbenzenes. Qualitative and semiquantitative results at parts per trillion levels (ppt or ng/l) can be obtained.[57]

Identification of various contaminants in the Niagara River water base–neutral fraction is shown in Fig. 8.8 and identified components are summarized in Table 8.6.[58]

8.4. POLYCYCLIC AROMATIC HYDROCARBONS

Polycyclic aromatic hydrocarbons (PAH) comprise a class of compounds with many isomeric series, having molecular structures based on fused benzene rings. In general, any organic compound containing two or more benzene rings may be considered a PAH. As a general class, PAH represent aromatic moieties ranging in molecular weight from 128 up to 500. These compounds may differ in number and position of fused rings and substituents and they may also contain various heteroatoms such as oxygen for polycyclic quinones, nitrogen in azaarenes, such as carbazoles and sulfur-containing polycyclic aromatic hydrocarbons, or thiaarenes (e.g., thiophenes).

8.4.1. General Considerations

PAH compounds and heterocyclic aromatic compounds (HAC) may undergo chemical reactions characteristic for aromatic compounds. It has been considered that the majority of PAH present in the ecosystem has been formed from low-, intermediate-, and high-temperature combustion, biosynthesis, and diagenesis.[59]

The sources of HAC and PAH compounds found in the atmospheric environment include natural and industrial emissions. Among the natural sources of PAH and HAC, the uncontrolled burning of wood, coal, and agricultural burning is considered to be an important route of entry into the atmospheric and aquatic environment. However, minerals composed of pure PAH and their analogs are well known, especially those associated with mercury ores. Pendletonite is pure coronene. Idrialite is a mixture of angularly annelated chrysene and picene along with alkylated analogs and nitrogen and sulfur analogs.[60] Industrial sources include municipal incineration, catalytic cracking of naphtha, coke production, and power generation stations where incomplete combustion of organic matter is performed.

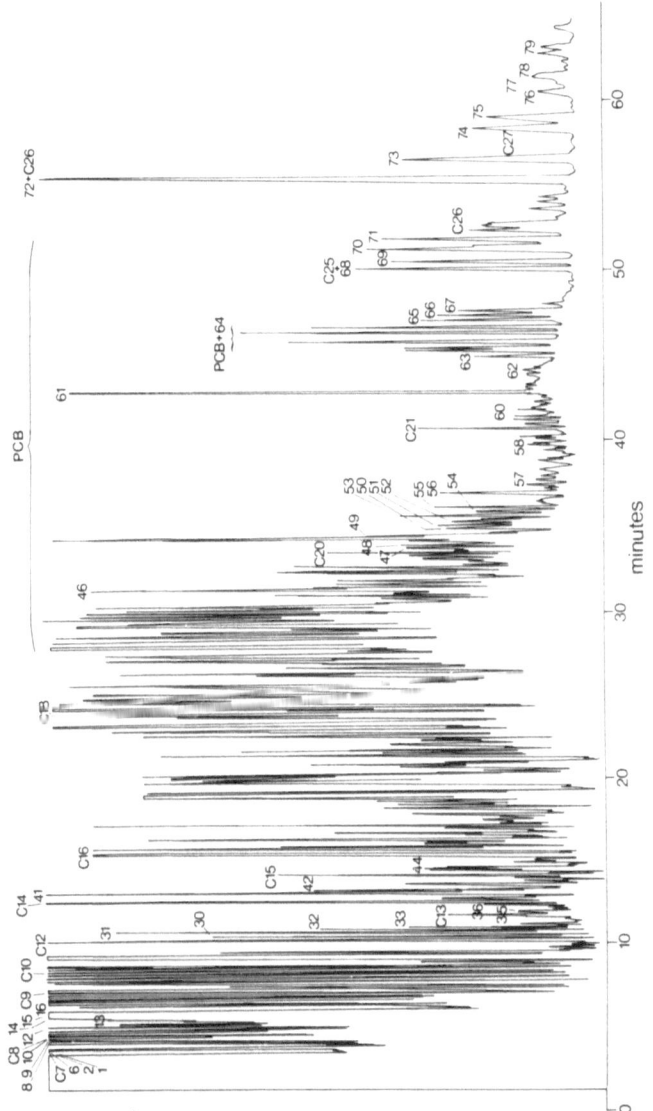

FIGURE 8.8. Base–neutral extract of the Niagara River water collected on the XAD-2 resin (total volume 480 l). Column: WCOT 30 m × 0.25-mm SE-54. Temperature programmed from 70 to 250 °C at 4 °C/min.

TABLE 8.6
Neutral Fraction Identification of Components

Peak	Substance	MW	b.p.	Ions m/z (relative intensity)
1	Methylcyclopentane	84	72	56(100); 55(30); 84(20)
2	Benzene	78	80	78(100); 77(15)
3	Cyclohexane	84	81	56(100); 84(72); 69(25)
4	n-Hexane	86	68	57(100); 56(55); 86(15)
5	Methyl cyclohexane	98	100	83(100; 98(47); 82(18)
6	Chlorobenzene	112	132	112(100); 114(31); 77(68)
7	n-Heptane	100	98	71(100); 57(90); 65; 70
8	Isooctane	114	99	71(100); 70(90); 85(32)
9	Toluene	92	111	91(100); 92(68); 93(5)
10	Methylheptane	114	118	1(100); 70; 55; 57
11	n-Octane	114	126	71; 85; 99; 114
12	Ethylbenzene	106	136	91(100); 106(30); 92; 77
13	p + m Xylenes	106	139	91(100); 106(62); 105(30)
14	Styrene	104	145	104(100); 103(40); 78(30)
15	o-Xylene	106	144	91(100); 106(60); 105(25)
16	Nonane	128	151	71; 85; 99; 128
17	Chlorotoluene	126	162	91(100); 126(42); 125(20)
18	Methylstyrene	118		118(100); 117(90); 91(30)
19	Mesitylene	120	165	105(100); 120(67); 119(16)
20	Dichlorobenzene	146	174	146(100); 148(60); 113(18)
21	Dichlorobenzene	146	179	146(100); 148(59); 113(16)
22	Indene	116	182	116(100); 115(85); 63; 89
23	Butylbenzene	134	183	119(100); 91(49); 134(20)
24	Decahydronaphthalene	138	195	138(100); 67(93); 96(90); 81
25	Divinylbenzene	130		130(100); 129(36); 128(33)
26	Tetrahydronaphthalene	132	208	104(100); 91(60); 132(45)
27	Naphthalene	128	211	128(100); 127(10); 129(9)
28	Dimethylindane	146		131(100); 51(20); 146(15)
29	Hexylbenzene	162		91(100); 92(72); 162(24)
30	Benzothiophene	134	221	134(100); 89(15); 135(10)
31	2-Methyl naphthalene	142	274	142(100); 141(65); 115(25)
32	1-Methyl naphthalene	142	245	142(100); 141(70); 115(25)
33	Triisopropyl benzene	204		189(100); 161(71); 204(24)
34	Tridecane	184	243	71; 85; 99; 113; (184)
35	Biphenyl	165	256	154(100); 153(35); 152(22)
36	Chloronaphthalene	162	263	162(100); 164(35); 127(30)
37	Dimethyl naphthalene	156	265	156(100); 141(95); 155(20)
38	Dimethyl phthalate	194	283	163(100); 77(25); 165(90); 194(10)
39	Trimethyl phosphate	170		170(100); 155(63); 169(24)
40	Dinitrotoluene	182	300	165(100); 89(64); 63(53)
41	Acenaphthene	154	278	154(97); 153(100); 152(45)

TABLE 8.6
(continued)

Peak	Substance	MW	b.p.	Ions m/z (relative intensity)
42	Dibenzofuran	168	287	168(100); 169(18); 139(18)
43	Pentadecane	212	271	71; 85; 99; 113; 127; 212
44	Fluorene	166	295	166(100); 165(91); 167(12)
45	Diethyl heptadecane	240	300	71; 85; 99; 113; 127; 240
46	Diethyl phthalate	222	345	149(100); 177(24); 150(10)
47	Octadecane	254	307	71; 85; 99; 113; 127(254)
48	9-Methyl fluorene	180	154[16]	165(100); 180(62); 178(19)
49	Dihydroanthracene	180	313	180(100); 179(62); 178(40)
50	2-Methyl fluorene	180		180(100); 165(95); 89(15)
51	1-Methyl fluorene	180		180(100); 165(96); 179(24)
52	Nonadecane	268	330	71; 85; 99; 113; 127; 268
53	Phenanthrene	178	340	178(100); 179(14); 176(18)
54	Anthracene	178	341	178(100); 179(14); 176(18)
55	Di-n-butyl phthalate	278	340	149(100); 150(9); 104(5)
56	2-Methyl anthracene	192	sub	192(100); 191(35); 189(20)
57	1-Methyl anthracene	192	200	192(100); 191(41); 189(20)
58	4,5-Dihydropyrene	204		202(100); 204(95); 203(88)
59	C_{20} eicosane	282	343	71; 85; 99; 113; 127; 282
60	Fluoranthene	202	384	202(100); 101(16); 100(11); 200(20)
61	Pyrene	202	394	202(100); 101(15); 200(22); 100(13)
62	Dimethyl anthracene	206		206(100); 205(22); 207(18)
63	C_{21} heneicosane	296	363	71; 85; 99; 113; 127; 296
64	Benzo[ghi]fluoranthene	226		226(100); 224(26); 227(24); 113(30)
65	C_{22} docosane	310	327	71; 85; 99; 113; 127
66	Benzo[a]fluorene	216	398	216(100); 215(72); 217(19)
67	C_{24} tetracosane	338	213[16]	43; 57; 71; 85; 99; 113; 141; 338
68	Methyl pyrene	216		216(100); 215(60); 217(18)
69	C_{25} pentacosane	350		69; 71; 83; 97; 111; 125; 350
70	Benzo[a]anthracene	228	438	228(100); 229(16); 226(16)
71	Chrysene/triphenylene	228	448	228(100); 226(18); 229(17)
	Heneicosane-C_{26}	364		69; 71; 83; 97; 111; 125; 364
72	Dioctyl phthalate	390	345	149(100); 279(11)
73	Benzo[k]fluoranthene	252	481	252(100); 253(22); 250(19)
74	Benzo[a]pyrene	252	496	252(100); 253(23); 250(22); 125(10)
	C_{29} nonacosane	378		69; 71; 83; 97; 111; 125; 378
75	Benzo[e]pyrene	252	493	252(100); 250(24); 253(22); 125(11)
76	Perylene	252	497	252(100); 253(22); 250(20)
77	Indeno[1,2,3,c,d]pyrene	276		276(100); 138(18); 277(24)
78	Dibenzo[a,h]anthracene	278		278(100); 139(39); 279(23)
79	Benzo[ghi]perylene	276		276(100); 138(20); 277(24)

It is recognized that PAH released to the atmospheric environment from combustion processes is associated with particulate matter. The PAH in atmospheric aerosols are present on fine particles and have significant residence times in the atmosphere.[61] Atmospheric particulate matter is deposited on the land or water via rainout or washout or involves gravitational settling. Industrial processes produce high levels of PAH that are discharged directly into the aquatic environment. It is evident that the industries producing high levels of aromatics are coke production, incinerators, refineries, shale oil processing, and acetylene production plants.

8.4.2. Sampling Requirements

Polycyclic aromatic hyrocarbons in air are generally adsorbed onto particulate matter. The air particulates are collected on glass filters in either a cascade impactor of a HiVol sampler.[62]

PAH are not very soluble in water (low $\mu g/l$). Sampling requires large volumes of water (50–100 l). Generally, PAH have been shown to be associated with suspended matter in waters, particularly in rivers or lakes that exhibit a relatively high turbidity. The initial problem when attempting to evaluate PAH in water is the concentration of other contaminants in comparison to the amount of PAH present.[63] Most analytical laboratories use single- or multiple-step extractions of water samples either with benzene, cyclohexane, or n-pentane.[64] A review covering the determination of PAH in water has been published.[65]

Sediment samples have usually been collected by grab samplers, and in some instances specially modified samplers have been used to collect the upper few centimeters of bottom sediments.[66] The initial sample size may vary as much as 10–1000 g. Sediment samples are usually air dried and ground. The samples are Soxhlet extracted with methanol for 24 hr; then benzene is added and the extraction continues for a further 24 hr. The hydrocarbons are then partitioned from the benzene–methanol extract into n-pentane. After removal of any extracted sulfur, the solution is subjected to gel permeation chromatography on a Sephadex LH-20 column. The fractions are then column chromatographed on alumina–silica gel to separate the PAH from the saturated hydrocarbons and olefins.[67]

Biological and wildlife samples for fish, birds, eggs, and oysters are usually hand collected or netted. The samples are wrapped in a pre-cleaned foil and placed in containers for freezing and shipment. Samples are homogenized in a suitable blender with a desiccant (usually sodium sulfate). Selected tissues from some birds are homogenized and freeze-dried prior to extraction. The protein-rich samples such as fish are saponified with alcoholic KOH (potassium hydroxide). Systems such as methanol–water–cyclohexane, N,N-dimethylformamide–water–cyclohexane, and column chromatography on Sephadex LH-20 are recommended.[68,71] PAH in oysters and mollusca are Soxhlet extracted by cyclohexane.[69,70] Cleanup and fractionation of the extracts is performed on a silica gel column.

8.4.3. High-Resolution WCOT Column Gas Chromatography

The similar chemical structures of many PAH compounds and their relatively high boiling points require high-resolution WCOT columns and thermally stable stationary phases for their chromatographic separation. Table 8.7 lists stationary phases suitable for separation of PAH.

The two most widely employed phases, SE-52 and SE-54, have nearly the same McReynolds' constants; however, for separating azaarenes, more polar phases such as OV-61 and SP-2340 look very promising.[72]

Examples of WCOT column applications are summarized in Table 8.8. High-temperature WCOT columns coated with SE-52, 15 m in length, can separate the PAH extracted from a carbon black having between 3 and 8 fused rings.[120] The high-performance fused-silica columns prepared by cross-linking of SE-54 show high temperature stability and low bleeding. They are suitable for the high-resolution separation at high column temperatures. They can even provide partial resolution between chrysene and triphenylene which are not separated using other stationary phases.[115]

In addition to column dimensions and type of liquid phase, the film thickness of a column as well as the carrier gas velocity, temperature-programming rate, and oven temperature are of great importance and should be optimized for the specific problem. Retention times, separation

TABLE 8.7
Stationary Phases Recommended for PAH Analyses

Stationary phase	McReynolds's constant
OV-1; methyl silicone gum	217
SP-2100; methyl silicone oil	229
SE-52; methyl, 5% phenyl silicone gum	334
SE-54; 1% vinyl, 5% phenyl, methyl silicone	337
OV-3; 10% phenyl methyl silicone	423

TABLE 8.8
Applications of WCOT Columns for the Determination of PAH

Application	Stationary phases	References
Air particulate matter	SE-52, SE-54, OV-1, Dexsil 300	75,76,80,97,98,99,100,101
Auto exhaust	OV-101, OV-17, Dexsil 300	79,102,103
Carbon black	SE-52	85
Cigarette smoke cond.	SE-52	76
Coal combustion products	SE-52	87
Coal liquid	SE-52, Dexsil 300 and 400	91,92,104,105
Coal gasification tar	SE-52, SE-54, OV-7, CX-20M	82,83,94,96,106,107
Dust	SE-52, SE-54	73,77
Industrial effluents	SE-54	95
Sediments	SE-52, SE-54	66,81,86,88,89,93
Marijuana smoke cond.	SE-52	84
Tobacco pyrolysis products	SE-54	108
Fish and shellfish	SE-52	69,71
Shale and lubricant oils	Dexsil 400, CP Sil5	109,116
Water	SE-52, OV-101, AP-L[a] Versamid	111,112,113,114
Atmospheric particulate	SE-52, CP Sil5, SE-30, OV-101	117,118
Standards	SE-52, SE-30, OV-3, XE-60, AP-L	74,75,77,78,110,119

[a] AP-L is Apiezon L.

efficiency, and capacity can be adjusted employing the correct film thickness. Comparison of the experimental findings are summarized in Figs. 8.9 and 8.10, showing dependence of column length, column internal diameter, and film thickness on chromatographic performance. The values presented in Table 8.9 indicate that the complexity of the chromatographic process in effective separation of PAH mixtures depends on many parameters that should be optimized.[86] The effect of various carrier gases is also a significant parameter. The elution temperatures are considerably lower using hydrogen as carrier gas than for nitrogen or helium. (Fig. 8.11).

8.4.4. Detection Systems

The most widely employed detector in PAH analysis is the flame ionization detector. This detector has excellent response, linearity, a wide dynamic range, and is suitable for quantitative analyses. Both flame ion-

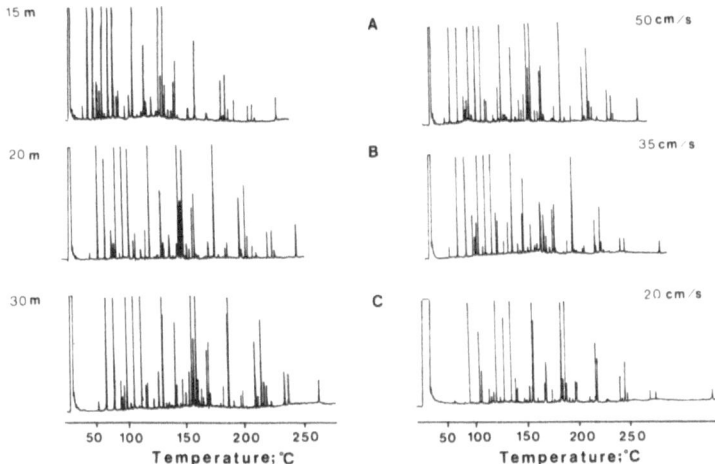

FIGURE 8.9. Coal tar extracts. Effect of length of the column and linear velocity (u_0) on speed of analysis. SE-52 WCOT columns; temperature programmed from 40 to 80 °C at 10 °C/min and then from 80 to 250 °C at 2 °C/min. (I) WCOT columns operated at $u_0 = $ 50 cm/sec; helium as a carrier gas; (II) WCOT columns 1 = 30-m × 0.3-mm-i.d. $u_{0.1} = $ 50 cm/sec; $u_{0B} = $ 35 cm/sec; and $u_{0C} = $ 20 cm/sec.

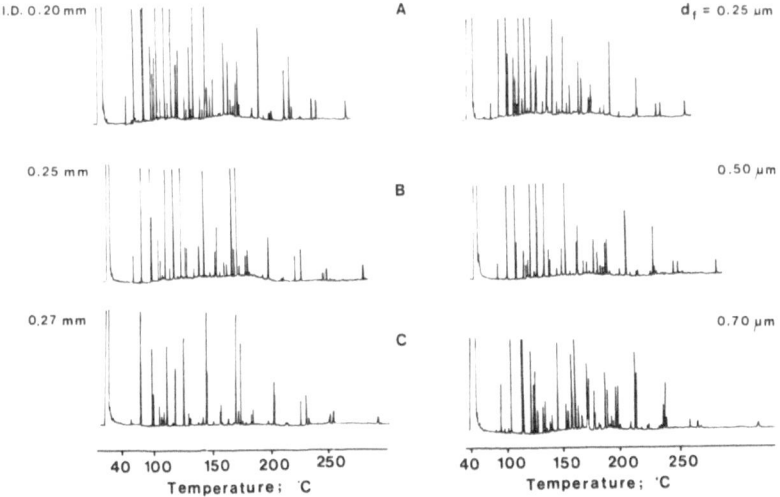

FIGURE 8.10. Coal tar extracts. Effects of column diameter and film thickness on speed of analysis. (I) All WCOT columns were 15 m long coated with SE-52; $u_0 = $ 50 cm/ sec. (II) WCOT columns were 15 m × 0.29-mm i.d., coated with SE-52.

TABLE 8.9
Summary of Data Presented in Figs. 8.10 and 8.11[a]

L (m)	i.d. (mm)	Elution temperature (°C)[b]	d_f (μm)	SN[c] capacity (ng)
15	0.30	198.4	0.1	60
20	0.30	207.2	0.1	60
30	0.30	218.8	0.1	65
15	0.20	228.8	0.25	50
15	0.25	220.0	0.25	80
15	0.27	217.2	0.25	90
15	0.29	206.0	0.15	80
15	0.29	217.2	0.25	100
15	0.29	223.6	0.35	120
15	0.29	229.6	0.50	130
15	0.29	238.8	0.70	150

[a] Helium was used as a carrier gas at 50 cm/sec.
[b] Elution temperature measured for benzo[a]pyrene.
[c] Separation number (SN) measured for the pair benzo[e]pyrene-perylene.
[d] Capacity of the WCOT column measured for phenanthrene.

ization and electron capture detectors have been evaluated for PAH analyses.[75,98] The ECD response differs from isomer to isomer, and it provides some means of differentiating among them. Of course, the most suitable and sensitive detection system is the mass spectrometer. Mass spectra of PAH exhibit very abundant molecular ions. On the other hand, the mass spectra of PAH isomers are in many cases indistinguishable when electron impact ionization is used. The chemical ionization mass

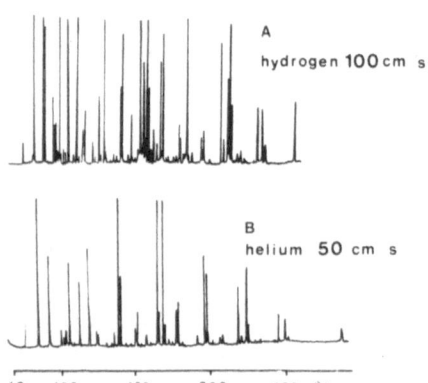

FIGURE 8.11. Coal tar extracts. Effect of a carrier gas on the speed of analysis. WCOT column 15-m × 0.27-mm-i.d. coated with SE-52, hydrogen u_0 = 100 cm/sec, helium u_0 = 50 cm/sec.

spectra of different isomers are in many cases different, and they can be used for diagnostic identification of the protonated molecular ion of a particular compound.[90]

8.4.5. Quantitative Analysis

Because of the superior qualitative information obtainable using WCOT columns over that from packed column or HPLC, there is an increasing demand for quantifying PAH compounds. The performance of injection systems is critical when analyzing mixtures of compounds of wide boiling range from different residues. In general, the on-column techniques are much more desirable than the splitless techniques or splitting.

For the analysis of Swiss river and lake sediments and particulate matter for PAH, both selected ion monitoring and conventional GC/MS has been employed.[89,121,122] PAH were identified in Zurich municipal sewage using GC/MS. PAH in Severn estuary sediments were determined also,[90] and levels of PAH industrial wastewaters using SIM GC/MS is also reported.[123] Analysis of PAH and organochlorine pollutants in Great Lakes herring gulls employing GC/MS and WCOT column gas chromatography has been described and confirmed using a SE-52 WCOT column.[124]

8.4.6. Qualitative Analysis

The relative retention, which presents retention data relative to a single arbitrary standard, and the retention index system using n-alkane references introduced by Kovats, are discussed in detail in Chapter 6. The Kovats retention index is the most widely accepted method for reporting chromatographic data. With well-defined chromatographic systems working at isothermal conditions, these systems can provide highly reproducible data. However, the degree of reproducibility is dependent on a uniformity of chemical composition of the stationary phase and degree of deactivation of the support material. The effect upon retention is greater when thin films of stationary phase are used. The active centra on the surface may interact with polar functional groups leading to an increased tailing. Since mixtures of PAH are exclusively analyzed by temperature programming, the Kovats retention index relationship developed for isothermal conditions must be modified.[125] The influence of the temperature-programming rate on the PAH retention index has been discussed[126] (Table 8.10).

TABLE 8.10[n]

PAH Retention Indices [a]

Compound	Index	Compound	Index
1,2-dihydronaphthalene	197.01	1-phenylnaphthalene	315.19
1,4-dihydronaphthalene	197.01	1,2,3,10b-tetrahydrofluoranthene	316.37
Tetralin	197.04	3-methylphenanthrene	319.46
Naphthalene	200.00	2-methylphenanthrene	320.17
Benzo[b]thiophene	201.47	2-methylanthracene	321.57
Indoline	204.74	o-Terphenyl	321.99
Indole	205.26	4H-cyclopental[def]phenanthrene	322.08
2-methylnaphthalene	218.14	9-methylphenanthrene	323.17
Azulene	219.95	4-methylphenanthrene	323.17
1-methylnaphthalene	221.04	1-methylanthracene	323.33
Biphenyl	233.96	1-methylphenanthrene	323.90
2,6-dimethylnaphthalene	237.58	9-n-butylfluorene	328.99
2,7-dimethylnaphthalene	237.71	9-methylanthracene	329.13
2-methylbiphenyl	238.77	Tetrahydropyrene	329.69
1,3-dimethylnaphthalene	240.25	2-phenylnaphthalene	332.59
1,7-dimethylnaphthalene	240.66	3,6-dimethylphenanthrene	337.38
1,6-dimethylnaphthalene	240.72	Hexahydropyrene	339.38
2,2'-dimethylbiphenyl	241.94	Fluoranthene	344.01
2,3-dimethylnaphthalene	243.55	9-isopropylphenanthene	345.78
1,4-dimethylnaphthalene	243.57	1,8-dimethylphenanthene	346.26
1,5-dimethylnaphthalene	244.98	9-n-hexylfluorene	348.54
Diphenylmethane	243.35	9-n-propylphenanthene	350.30
Acenaphthylene	244.63	Pyrene	351.22
1,2-dimethylnaphthalene	246.49	9-methyl-10-ethyl phenanthrene	359.91
1,8-dimethylnaphthalene	249.52	p-Terphenyl	366.10
2-ethylbiphenyl	250.85	Benzo[a]fluorene	366.74
Acenaphthene	251.29	11-methylbenzo[a]fluorene	367.04
4-methylbiphenyl	254.71	9,10-diethylphenanthrene	367.97
3-methylbiphenyl	254.81	1-methylpyrene	373.55
Dibenzofuran	257.17	2,7-dimethylpyrene	386.34
2,3,6-trimethylnapthalene	263.31	Dodecahydrotriphenylene	386.36
1-methylacenaphthylene	265.24	1,1'-binaphthyl	388.38
2,3,5-trimethylnaphthalene	265.90	Benzo[ghi]fluoranthene	389.60
Fluorene	268.17	Benzo[c]phenanthrene	391.39
3,3'-dimethylbiphenyl	271.87	9-phenylanthracene	396.38
9-methylfluorene	272.38	Benzo[a]anthracene	398.50
4,4'-dimethylbiphenyl	274.59	Chrysene	400.00
9,10-dihydroanthracene	284.89	Triphenylene	400.00
9-ethylfuorene	284.99	7-benz[de]anthrene	406.54
9,10-dihydrophenanthrene	287.09	9-phenylphenanthrene	406.90
Octahydroanthracene	287.69	Naphthacene	408.30
2-methylfluorene	288.21	2-methylbenz[a]anthracene	413.78
1-methylfluorene	289.03	1-methylbenz[a]anthracene	414.37
Octahydrophenanthrene	292.03	1-methyltriphenylene	416.32
1,2,3,4-tetrahydrophenanthrene	297.21	9-methylbenz[a]anthracene	416.50
Phenanthrene	300.00	3-methylbenz[a]anthracene	416.63
Anthracene	301.69	6-methylbenz[a]anthracene	417.57

TABLE 8.10ᵃ

(continued)

Compound	Index	Compound	Index
3-methylchrysene	418.10	Benzo[k]fluoranthene	442.56
2-methylchrysene	418.80	1,6,11-trimethyltriphenylene	446.24
12-methylbenz[a]anthracene	419.39	Benzo[e]pyrene	450.73
5-methylchrysene	419.68	Benzo[a]pyrene	453.44
1-phenylphenanthrene	421.66	Perylene	456.22
1-methylchrysene	422.87	1,3,6,11-tetramethyltriphenylene	461.72
o-Quaterphenyl	423.63	3-methylcholanthrene	468.44
1,3-dimethyltriphenylene	432.32	m-Quaterphenyl	472.81
1,12-dimethylbenz[a]anthracene	436.82	Indeno[1,2,3-cd]pyrene	481.87
Benzo[j]fluoranthene	440.92	p-Quaterphenyl	488.18
Benzo[b]fluoranthene	441.74	Dibenzo[ac]anthracene	495.01
		Benzo[ghi]perylene	501.32

ᵃFrom Reference 119.

8.5. HETEROCYCLIC AROMATIC HYDROCARBONS

A separate subgroup of PAH are heterocyclic aromatic hydrocarbons (HAC). They can be divided into three categories as follows:

(i) Oxygen-containing PAH polycyclic quinones;

(ii) Nitrogen-containing PAH neutral substances; e.g., carbazoles and acridines (azaarenes); basic nitrogen-containing PAH such as amino PAH; acidic nitrogen-containing PAH such as nitro PAH; and

(iii) Sulfur-containing PAH known as thiaarenes, e.g., thiophenes, benzothiophenes, etc.

All heterocyclic aromatic hydrocarbons have come under increasing scrutiny as potential contaminants in our ecosystem.

8.5.1. Nitrogen-Containing PAH (Azaarenes)

The largest industrial point source of atmospheric PAH and HAC is the emission arising from the production of coke. These compounds are found in crude oils and contain from two to seven fused rings, which mostly contain single nitrogen atoms in their molecules.[127] As a possible source of pollution, they are found in tobacco smoke,[128] air particulate matter,[129] and coal.[130] The accepted nomenclature of heterocyclic aromatic hydrocarbons is that adopted by IUPAC. Some examples of the nomenclature and structural configurations are given in Table 8.11.

Alkylsubstituted quinolines and benzoquinolines are recognized as major nitrogen-containing bases of crude oil. The lack of reference com-

TABLE 8.11
Molecular Weights and Structures of Some Heterocyclic Aromatic Compounds

Compound	Mol. wt.	Molecular structure
Quinoline	129	
Acridine	179	
Benz[a]acridine	229	
Dibenz[a,j]acridine	279	
1-azafluoranthene	203	
4-azapyrene	203	
Carbazole	167	
Benzo[a]carbazole	217	
Dibenzo[c,g]carbazole	267	
9,10-anthraquinone	208	
1,6-pyrenenquinone	232	
Aminonaphthalene	143	
2-nitronaphthalene	173	

pounds is partly responsible for difficulties of the location of the nitrogen atom in triameric azaarenes.[131] Because of the nature of crude oil basic extracts, no single analytical technique allows the identification of the position of the heteroatom in individual HAC. Even GC/MS cannot distinguish among them. WCOT column gas chromatography provides the best selectivity for isomers differing in the location of the nitrogen atom and at the same time produces minimum interferences between alkylated benzoquinolines. Liquid phase films of OV-61 ($d_f = 0.1$–0.2 μm) can separate a wide range of alkylquinolines together with several ring isomers of benzoquinolines and alkylated aromatic amines.[125] Retention indices of different azaarenes are given in Table 8.12.

TABLE 8.12
Retention Indices of Azaareness[a]

Compound	Index	Compound	Index
Quinoline	210.26	Benzo[def]carbazole	363.92
Isoquinoline	214.14	Benz[c]acridine	392.60
1-methylindole	216.90	Benzo[a]acridine	398.65
Indole	222.66	1-azabenz]a]anthracene	400.00
7-azaindole	223.70	Azachrysene	401.16
2-methylquinoline	224.13	Benzo[a]carbazole	402.22
8-methylquinoline	225.18	1-azachrysene	407.18
1-methylisoquinoline	229.21	4-azalfuorene	297.85
7-methylquinoline	231.37	Phenazine	294.37
3-methylquinoline	232.47	Benzo[h]quinoline	301.94
7-methylindole	235.49	Benzo[f]quinoline	307.94
4-methylquinoline	235.77	Phenatridine	307.94
3-methylindole	239.20	Carbazole	311.71
2-methylindole	240.10	3-methylbenzo[f]quinoline	320.26
2,7-dimethylquinoline	244.04	2-methylbenzo[f]quinoline	320.50
2,6-dimethylquinoline	244.19	2-methylacridine	324.34
1,2-dimethylindole	244.42	Dibenz[a,c]phenazine	474.08
2,4-dimethylquinoline	247.96	Dibenz[a,h]acridine	488.55
Azabiphenyl	252.35	Dibenz[a,i]carbazole	490.57
2,5-dimethylindole	256.65	Dibenz[a,j]acridine	490.66
2,3-dimethylindole	257.32	Dibenzo[a,g]carbazole	502.30
1-methylacridine	324.45	Dibenzo[c,g]carbazole	502.92
3-methylcarbazole	328.81	1-cyanonaphthalene	256.75
2-methylcarbazole	329.61	2-cyanoanthracene	260.88
9-methylacridine	331.15	9-cyanoanthracene	350.46
4-methylcarbazole	331.88	9-cyanophenanthrene	351.84
1,4-dimethylcarbazole	318.10	Benzo[b]carbazole	409.63
1,2-dimethylcarbazole	347.31	2-azachrysene	411.49
2-azafluoranthene	347.39	Benzo[c]carbazole	411.89
1-azafluoranthene	348.17	2,2'-biquinoline	422.56
1,3-dimethylcarbazole	348.45	7,9-dimethylbenz[c]acridine	438.32
7-azafluoranthene	350.50	5,7-dimethylbenz[a]acridine	438.38
1-azapyrene	357.73	7,10-dimethylbenz[a]acridine	439.46
4-azapyrene	357.94	10-azabenzo[a]pyrene	455.40
2-azapyrene	362.43	9,10,12-trimethylbenz[a]acridine	466.79

[a]From Reference 131.

8.5.2. Polycyclic Aromatic Amines (PAA)

Polycyclic aromatic hydrocarbons containing an amino group were identified in cigarette smoke[134] and in synthetic fuels,[132] where they have become of increased concern because of the mutagenicity and carcinogenicity exhibited by this class of compounds. Their occurrence in coal-derived liquids and solvent-refined coal has been reported.[135]

Anilines are the major basic nitrogen-containing compounds of the light end product and substantial components of the digester condensate.

With the exception of some aminobiphenyls, no higher aromatic primary amines were identified. Significant amounts of diphenylamines and aminobiphenyls were found in the coal extract solution. Their origin is traced to the ring opening of carbazole-type structures during extraction according to:

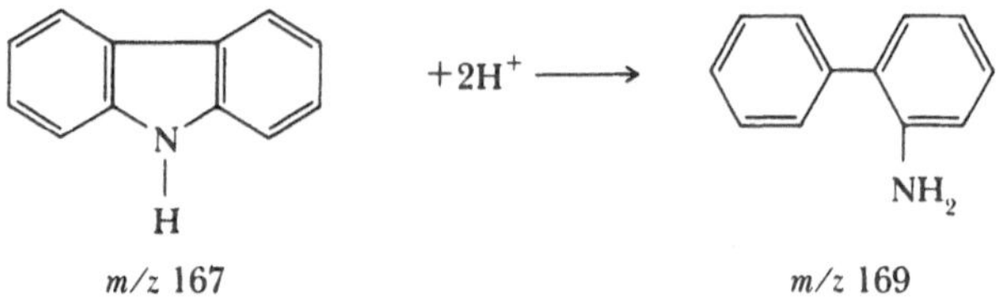

$+2H^+ \longrightarrow$

m/z 167   *m/z* 169

Neutral   Basic

8.5.2a. Isolation of Polycyclic Aromatic Amines. Solid and liquid samples are dissolved in dichloromethane and extracted with hydrochloric acid or H_2SO_4 for 24 hr.[131] The acid extracts are combined, filtered, and then neutralized to pH 12–14 with NaOH. The bases are extracted with diethyl ether or dichloromethane and preconcentrated to a small volume under nitrogen or vacuum at room temperature before analysis.

8.5.2b. Derivatization of Polycyclic Aromatic Amines. A suitable derivatization technique such as methylating reagent dimethylformamide dimethylacetal (Methyl-8)[133] and fluoroacetylation,[132] which selectively reacts with polycyclic aromatic amine (PAA), helps to distinguish between free amines and their methyl azaarene homologues. Reactions and reaction products formed are presented in the following reaction scheme:

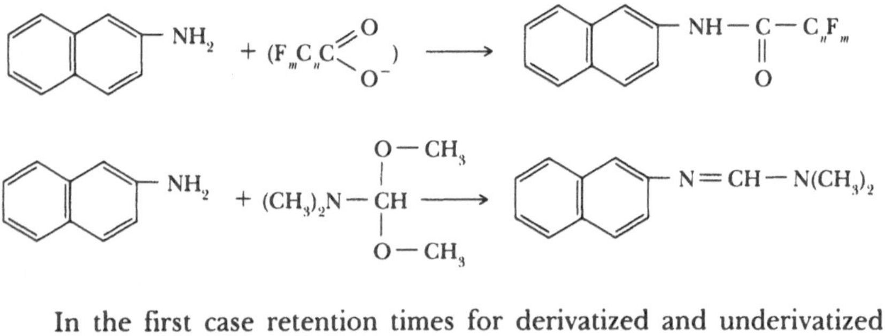

In the first case retention times for derivatized and underivatized amines do not differ too much. However, when Methyl-8 is used, derivatized primary amines have considerably longer retention times than the unreacted amines. The derivatization of PAA to either the corresponding trifluoracetyl-(TFA) or pentafluoropropionyl-amides or *N*-dimethylaminomethylene derivatives provide a solution to identify and quantify individual amine compounds.

8.5.2c. Analysis of Polycyclic Aromatic Amines. Selective detection of fluorinated molecules with ECD and also by GC/MS takes full advantage of the properties conferred by the derivatization to ease quantitation of individual components.

GC/MS confirmation indicates that the $(M-97)^+$ ion is formed, and it corresponds to the loss of the TFA group and provides simultaneous information of the $(M-1)^+$ ion of the underivatized aromatic amine. Also a diagnostic ion in the mass spectrum results from the loss of the HCN molecule from the $(M-97)^+$ moiety. Similar fragmentation patterns are observed when pentafluoropropionyl (PFP) derivatives are analyzed by GC/MS. The major fragment ion corresponding to the loss of the PFP group as the $(M-147)^+$ ion is observed in the mass spectrum.

A typical ECD WCOT column gas chromatogram of the underivatized and derivatized PAA of solvent refined coal is shown in Fig. 8.12. Retention indices for PAA are given in Table 8.13.

The WCOT columns employed in the separation of various NAC are usually coated with SE-54, OV-73 (5.5% phenyl, 94.5% methylpolysiloxane gum), or DB-5 fused-silica WCOT columns. WCOT columns are 25–40 m in length, having internal diameters between 0.25 to 0.34 mm and film thicknesses between 0.05/0.25 μm.

8.5.3. Nitro Polycyclic Aromatic Hydrocarbons

Nitro polycyclic aromatic hydrocarbons (NAC) can be formed in the environment by reaction between nitrogen oxides and PAH under sunlight-catalyzed conditions. They have been detected in several environmental samples such as airborne particulate matter,[136,137] diesel exhaust particles,[138] and carbon black.[139,140] These studies suggest that nitro-PAH are widespread in the environment at lower concentrations than PAH.

A chromatogram of NAC standard mixture is shown in Fig. 8.13 using FID. Individual NAC show approximately 20–70% relative responses (relative to benzo[e]pyrene) for the FID. ECD can also be used for detecting and quantitation of NAC.[140] Their relative retention times are given in Table 8.14.

By use of negative ion chemical ionization GC/MS, with CH_4 as the reactant gas and selective ion monitoring, the detection limit is in the low picogram level. It seems that NAC mass spectra do not exhibit significant dependence on pressure and temperature.

For high-boiling NAC the use of cold on-column injection and high temperature stability for stationary phases are required. The use of se-

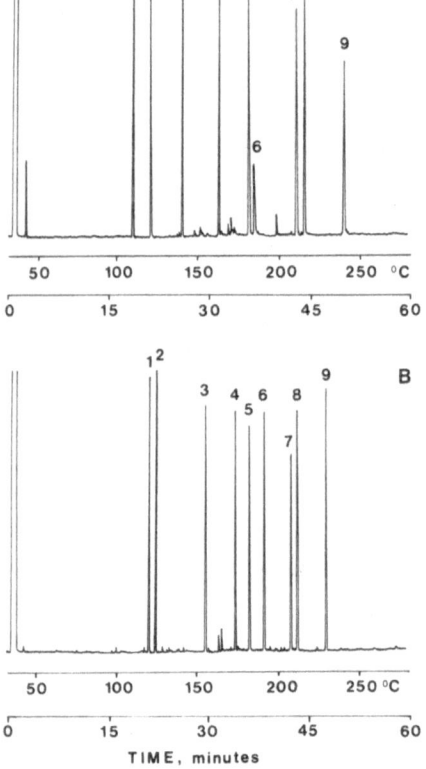

FIGURE 8.12. Chromatogram of (A) underivatized and (B) PFP-derivatized polycyclic aromatic amines on a 30 m × 0.30-mm-i.d. fused-silica WCOT column coated with SE-52. Temperature programmed from 40 to 250 °C at 4 °C/min. Peaks: (1) 1-aminonaphthalene; (2) 2-aminobiphenyl; (3) 4-aminobiphenyl; (4) 2-aminofluorene; (5) 1-aminoanthracene and 9-aminophenanthrene; (6) 2-aminoanthracene; (7) 3-aminofluoranthene; (8) 1-aminopyrene; (9) 6-aminochrysene.

lective detection systems such as nitrogen–phosphorus (NPD), ECD, and mass spectrometric detection systems are preferred to the FID. Retention indices of some NAC are presented in Table 8.15.

8.5.4. Sulfur-Containing Polycyclic Aromatic Hydrocarbons

First reports covering heterocyclic sulfur-containing (HSC) compounds in the environment appeared in 1973.[141] HSC have been identified and confirmed using GC/MS–SIM in shellfish and fish.[142-144] The methodologies are sufficiently described for isolating HSC from diverse environmental samples such as for fish, shellfish, sediment, and air particulate matter.

Several recent studies support the observation that the sulfur-containing PAH may be the most persistent of all the PAH in the environ-

TABLE 8.13
Retention Indices of Polycyclic Aromatic Amines[a]

Compound	Index	Compound	Index
1-aminoindan	207.63	9-aminoanthracene	363.91
5-aminoindan	232.12	3-aminophenanthrene	365.60
1-aminonaphthalene	262.98	2-aminophenanthrene	365.80
2-aminonaphthalene	265.53	2-aminoanthracene	367.45
2-aminobiphenyl	273.63	3-aminofluoranthene	409.97
3-methyl-2-aminonaphthalene	283.73	4-aminopyrene	412.31
4-aminobiphenyl	298.05	2-aminopyrene	413.83
4-aminofluorene	325.11	1-aminopyrene	415.39
1-aminofluorene	327.21	2-aminobenzo[c]phenanthrene	450.10
3-aminofluorene	329.08	4-aminobenzo[c]phenanthrene	451.51
2-aminofluorene	331.91	6-aminochrysene	463.19
4-aminophenanthrene	353.97	5-aminochrysene	487.88
1-aminophenanthrene	362.62	7-aminobenzo[a]pyrene	511.98
9-aminophenanthrene	362.83	6-aminobenzo[a]pyrene	515.66

[a]From Reference 135.

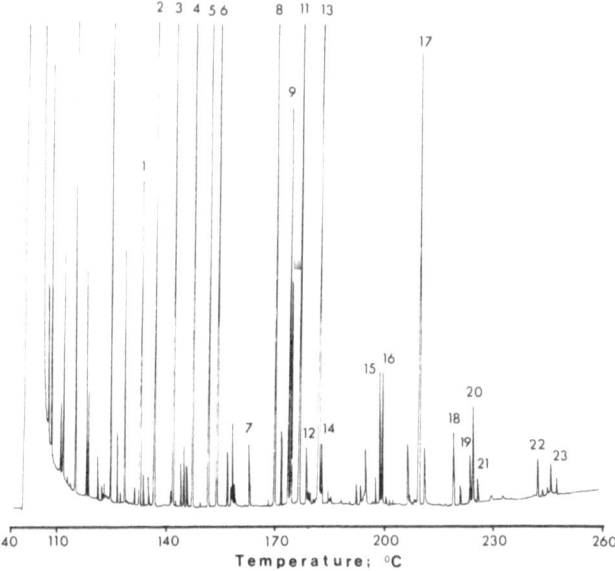

FIGURE 8.13. Standard mixture of nitro-heterocyclic aromatic hydrocarbons in a mixture of a coal tar sample. WCOT column 29 m × 0.32-mm i.d. coated with OV-73 (d_f = 0.05 μm). Temperature programmed from 40 to 260 °C at 3 °C/min. Peaks: (1) Fluorene; (2) methyl-1-nitronaphthalene; (3) 2-nitrobiphenyl; (4) phenanthrene; (5) 3-nitrobiphenyl; (6) 4-nitrobiphenyl; (7) anthraquinone; (8) fluoranthene; (9) pyrene; (10) 2-nitrofluorene; (11) 9-nitroanthracene; (12) benzo[a]fluorene; (13) 2-nitrofluorenone; (14) benzo[b]fluorene; (15) benzo[a]anthracene; 16 chrysene + triphenylene; (17) 1-nitropyrene; (18) benzo[b]-fluoranthene; (19) benzo[e]pyrene; (20) benzo[a]pyrene; (21) perylene; (22) indeno[1,2,3-cd]-pyrene; (23) benzo[ghi]perylene.

TABLE 8.14

Nitroaromatic Compounds, Their Characteristic Ions and Relative
Retention Times on OV-73 Stationary Phase[a]

Compound	m/z	RRT
2-methyl-1-nitronaphthalene	187	0.474
2-nitrobiphenyl	199	0.514
3-nitrobiphenyl	199	0.594
4-nitrobiphenyl	199	0.611
2-nitrofluorene	211	0.755
9-nitroanthracene	223	0.773
2-nitrofluorenone	225	0.809
4-nitropyrene	247	0.981
1-nitropyrene	247	1.000
2-nitropyrene	247	1.010
1,3-dinitropyrene	292	1.157
1,6-dinitropyrene	292	1.180
1,8-dinitropyrene	292	1.196
1,3,6-trinitropyrene	337	1.302
1,3,6,8-tetranitropyrene	382	1.388

[a] Column: WCOT 29 m × 0.32-mm i.d. coated with OV-73, $d_f = 0.05$ μm;
carrier gas, helium; flow rate 40 cm/sec at 100 °C; column temperature 100–
325 °C at 5 °C/min.

ment.[145,146] They are significantly more toxic than either the PAH or
nitrogen-containing HAC.

The proposed fractionation schemes for preparing fish, sediment,
and airborne particulate matter samples are based on an alumina column
adsorption chromatographic cleanup. Airborne particulate matter samples (15 g) are Soxhlet extracted with 1 : 3 methanol–benzene (250 ml)
for 12 hr. The extract is preconcentrated to 5 ml using a rotary evaporator.
One milliliter aliquot is applied to 3 g of neutral aluminum oxide (activity
1, 80–100 mesh) column and transferred to the top of a 6-g alumina
column. The sample is eluted with 100 ml of chloroform containing
0.75% ethanol. The eluate is reduced in volume to 1 ml and transferred
to a 22-mm-i.d. column packed with Bio-Beads SX-12 (60 g) and eluted
with methylene chloride. The first portion (55 ml) contains aliphatics and

TABLE 8.15

Retention Indices of Some Nitro Aromatic Compounds

Compound	Index	Compound	Index
5-nitroindan	261.55	4-nitrobiphenyl	314.59
1-nitronaphthalene	274.95	2-nitrofluorene	353.06
2-nitroanaphthalene	280.63	9-nitroanthracene	357.42
2-nitrobiphenyl	290.25	1-nitropyrene	421.48
3-nitrobiphenyl	310.09	6-nitrobenzo[a]pyrene	501.71

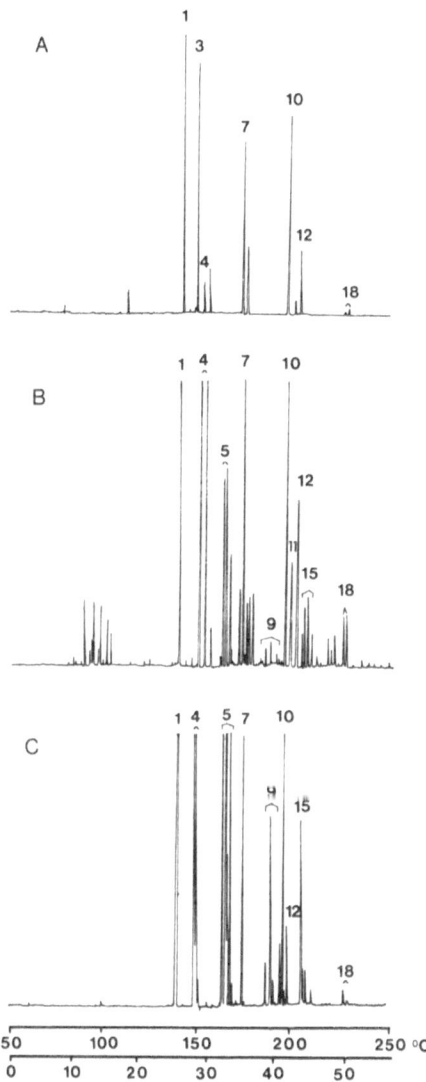

FIGURE 8.14. Chromatograms of sulfur-containing heterocyclic aromatic hydrocarbons in (A) Black River sediment sample, (B) Washington air particulate matter, and (C) Sacramento River striped bass. Temperature programmed from 50 to 265 °C at 4 °C/min using FPD. Peaks. (1) dibenzothiophene; (3) naphtho[2,3-*b*]thiophene; (4) methyl dibenzothiophenes; (5) C_2-dibenzothiophenes; (7) phenanthro[4,5-*bcd*]thiophene; (9) methyl phenanthro-[4,5-*bcd*]thiophenes; (10) benzo[*b*]naphtho[2,1-*d*]thiophene; (11) benzo[*b*]-naphtho[1,2-*d*]thiophene; (12) benzo[*b*]-naphtho[2,3-*d*]thiophene; (15) methyl 4-ring thiophenes; (18) peri-condensed 5-ring thiophenes.

the next 100 ml contains PAH and HACs. Two-step alumina fractionation (after preconcentration of the 100 ml volume to 1 ml) with benzene and chloroform produces relatively clean fractions suitable for HRGC and GC/MS.

Fish and sediment samples are analyzed according to the same fractionation scheme. It is evident that the common step in both schemes is the alumina column separation step of the PAH fraction into specific

TABLE 8.16
Compounds Identified in Sediment, Air, and Fish Samples Shown in
Fig. 8.14

Peak	Sulfur-containing compound
1	Dibenzothiophene
3	Naphto[2,3-b]thiophene
4	Methyldibenzothiophenes
5	C-2-dibenzothiophenes
7	Phenanthro[4,5-bcd]thiophene
9	Methylphenanthro[4,5,-bcd]thiophenes
10	Benzo[b]naphtho[2,1,-d]thiophene
11	Benzo[b]naphtho[1,2,-d]thiophene
12	Benzo[b]naphtho[2,3-d]thiophene
15	Methyl 4-ring thiophenes
18	Pericondensed 5-ring thiophenes

chemical classes. The base hydrolysis, aqueous wash, and gel permeation steps are essential for separating lipids from PAH which are present in the fish tissue.

Figure 8.14 shows representative chromatograms of the PAH and sulfur-containing HAC fractions isolated from air particulates, sediment, and fish tissues using a 20-m $\times$ 0.3-mm-i.d. fused-silica WCOT column coated with SE-52 ($d_f = 0.34$ μm). Compounds are identified in Table 8.16. Retention indices for sulfur-containing HAC are given in Table 8.17.

8.6. HALOGENATED ALIPHATIC HYDROCARBONS

The environmental applications of WCOT column GC and GC/MS for environmental contaminants have become increasingly important. They can be used both as an excellent separation technique and a sensitive detection system for polyhalogenated xenobiotic chemicals, providing simultaneously confirmation of their structure. A review of polyhalogenated chemical residues identified in fish tissue contains over 85 references.[147]

Low molecular weight halogenated hydrocarbons have long been used as industrial solvents, aerosol propellants, and polymer precursors. The development of trapping techniques for volatile compounds in connection with GC/MS analysis has confirmed the ubiquity of these pollutants. There is an increasing concern about the presence of these pollutants in water, sediment, fish, biota, and air. Because of their lipophilic nature, there is also concern about the extent to which they will accumulate in the fatty tissues of animal life.

TABLE 8.17
Retention Indices of Sulfur-Containing HAC[a]

Compound	I	Compound	I
Benzo[b]thiophene	201.57	Benzo[1,2]phenaleno[3,4-bc]thiophene	447.66
1,2,3,4,4a,4b,-hexahydrodibenzothiophene	271.69	Triphenyleno[4,5-bcd]thiophene	448.45
Naphtho[1,2-b]thiophene	295.80	Pyrene[1,2-b]thiophene	449.30
Dibenzothiophene	296.01	Chryseno[4,5-bcd]thiophene	450.62
Naphtho[2,1-b]thiophene	300.00	Pyreno[2,1-b]thiophene	455.01
Naphtho[2,3-b]thiophene	304.47	Benzo[4,5]phenaleno[1,9-bc]thiophene	455.99
Phenanthro[4,5-bcd]thiophene	348.75	Benzo[4,5]phenaleno[9,1-bc]thiophene	457.30
Phenanthro[6,7-bc]thiophene	353.45	Benzo[b]phenanthro[4,3-d]thiophene	470.47
Benzo[b]naphtho[2,1-d]thiophene	389.37	Dinaphtho[2,1-b; 1',2'-d]thiophene	472.62
Benzo[b]naphtho[1,2-d]thiophene	392.92	Dinaphtho[1,2-b; 2',1'-d]thiophene	482.60
Phenanthro[9,10-b]thiophene	394.96	Benzo[1,2]phenanthro[4,3-bc]thiophene	482.99
Phenanthro[4,3-b]thiophene	395.03	Dinaphtho[1,2-b; 1',2'-d]thiophene	486.58
Anthra[1,2-b]thiophene	395.39	Benzo[b]phenanthro[9,10-d]thiophene	487.32
Phenanthro[1,2-b]thiophene	396.01	Benzo[b]phenanthro[3,4-d]thiophene	487.76
Phenanthro[3,4-b]thiophene	396.43	Anthra[1,2-b]benzo[d]thiophene	488.45
Anthra[2,1-b]thiophene	399.31	Benzo[b]phenanthro[2,1-d]thiophene	488.89
Phenanthro[2,1-b]thiophene	400.59	Dinaphtho[1,2-b; 2',3'-d]thiophene	489.14
Phenanthro[3,2-b]thiophene	401.89	9,13-H-triphenyleno[2,3-b]thiophene	489.02
Phenanthro[2,3-b]thiophene	402.19	Benzo[b]phenanthro[1,2-d]thiophene	492.31
Anthra[2,3-b]thiophene	407.57	Benzo[b]phenanthro[2,3-d]thiophene	493.31
2-[2'-naphthyl]benzo[b]thiophene	430.65	Triphenyleno[1,2-b]thiophene	494.41
Benzo[2,3]phenanthro[4,5-bcd]thiophene	443.29	Triphenyleno[2,3-b]thiophene	500.00
Pyrene[4,5-b]thiophene	446.51		

[a] From Reference 145.

8.6.1. Halogenated Alkanes and Alkenes

Halogenated methanes and other halogenated alkanes and alkenes were discussed previously under the priority pollutants (Sec. 8.2).

8.6.2. Chlorinated Dimethanonaphthalene and Methanoindene Pesticides

Organochlorine pesticides are widely used in our environment, and it is well recognized that under environmental conditions complex multicomponent mixture of pesticides and their degradation and photoalteration products will occur. Chlorinated dimethanonaphthalenes and methanoindene pesticides such as Aldrin, (hexachloro-hexahydro-endo-exo-dimethanonaphthalene), Dieldrin (hexachloroepoxy-octahydro-endo-exo-dimethanonaphthalene), Endrin (hexachloroepoxy-octahydro-endo-endo-dimethanonaphthalene) are the cyclodiene class of chlorinated hydrocarbon insecticides. Chlordane (octachloro-hexahydro-4,7-

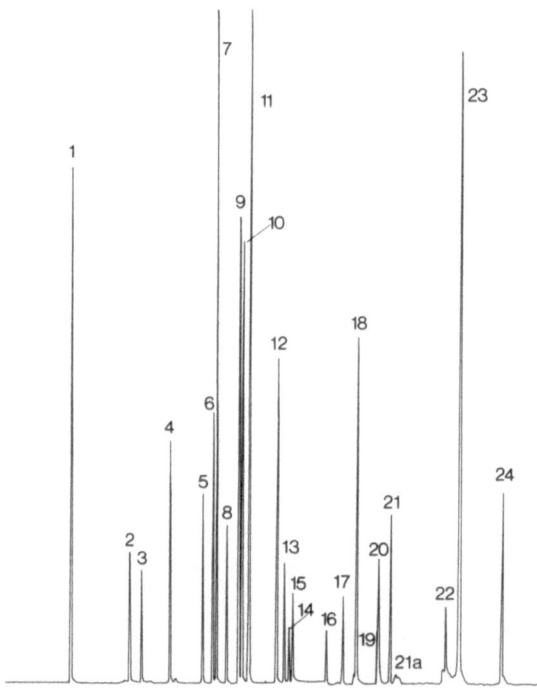

FIGURE 8.15. Chromatogram of the dimethanonaphthalene and methanoindene pesticides and their photo-isomers on OV-101 (20 m × 0.25-mm i.d.) WCOT column conditions: 80–250 °C, 3-min hold, temperature programming rate 4 °C/min; peaks are identified in Table 8.18.

TABLE 8.18
Organochlorine Pesticides Analysis

Peak	Name	Peak	Name
1	Chlordene	13	Endrin
2	Heptachlor	14	Dihydromirex
3	Aldrin	15	p,p'-DDE
4	Telodrin	16	Monohydromirex
5	Oxychlordane	17	Kepone
6	Photo-heptachlor	18	Photoaldrin
7	Photo-chlordene	19	p,p'-DDT
8	Heptachloro epoxide	20	Mirex
9	trans-Chlordane	21	Endrin aldehyde
10	trans-Nonachlor	22	Endrin ketone
11	cis-Chlordane	23	Methoxychlor
12	Dieldrin	24	Photo-dieldrin

methanoindene), heptachlor (heptachloro-tetrahydro-4,7-methanoin-
dene), and heptachlor epoxide (heptachloro-tetrahydro-2,3-epoxy-4,7-
methanoindene) are technical products containing a mixture of related
compounds. Figure 8.15 shows a chromatogram of a standard mixture
of various dimethanonaphthalene and methanoindene compounds and
some of their metabolites. Individual peaks are identified in Table 8.18.

A simple multimatrix method for the isolation of δ-hexachlorocy-
clohexane and some chlorinated insecticides is described by Steinwandter.[148] The quantitative separation of active substances such as HCB,
heptachlor, aldrin, heptachlor epoxide, α- and γ-chlordane, dieldrin, endrin, and DDT compounds is achieved by silica gel cleanup. The elution
is performed with petroleum ether–methylene chloride (80–20). Under
these conditions α-HCH and the chlorinated pesticides are eluted quantitatively. The silica gel column has a capacity to retain fat up to 500 mg.
Quantitative determination of the 17 pesticides is performed by WCOT
column GC on OV-17 and ECD, and GC/MS detection is employed. The
recoveries of γ-HCH and other chlorinated pesticides from the fatty tissue
and vegetables are between 95 to 100%.

The hexachlorocyclopentadiene-derived pesticides such as technical
chlordane, trans- and cis-nonachlor, and chlordene contain many different
components. Technical grade chlordane contains up to 45 different impurities.[149]

Hexachloro-1,3-butadiene is a by-product from the synthesis of
perchloroethylene. It has been found as a residue in fish and water from
the lower Mississippi River[150,151] and the Niagara River.[10,152] The
1,2,3,4,7,7-hexachloro-2,5-norbornadiene, 1,2,3,4,5-endo-7,7-heptach-
loro-2-norbornene, and 1,2,3,4,7,7-hexachloro-5,6-endo-epoxi-2-nor-

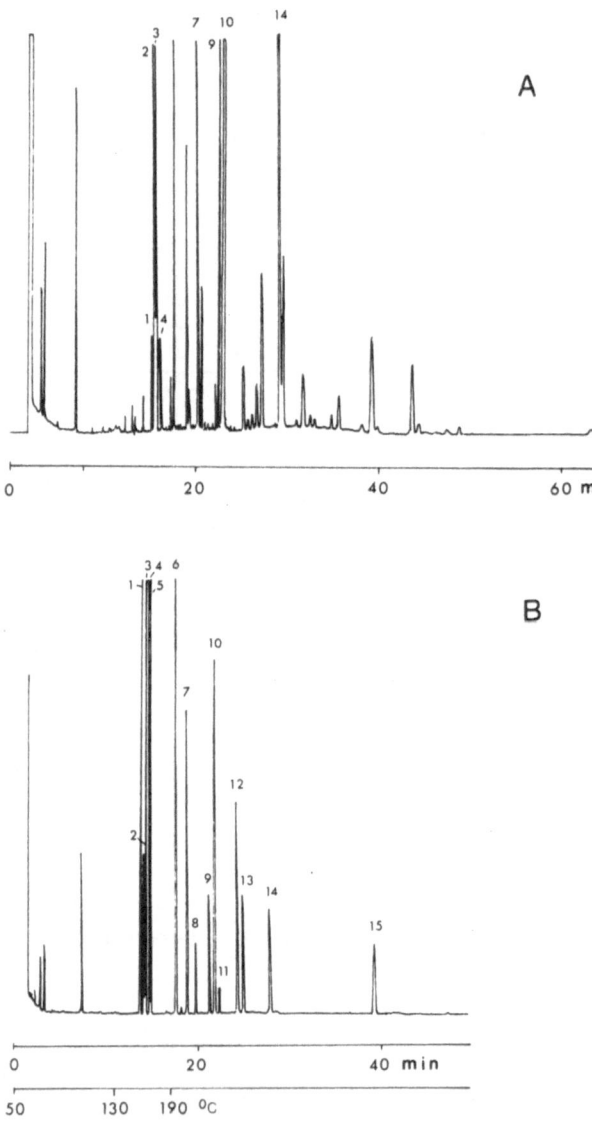

FIGURE 8.16. Organochlorine compounds and PCB residue in (A) human bone marrow and (B) standard. Peaks: (1) α-HCH; (2) β-HCH; (3) HCH; (4) γ-HCH; (5) δ-HCH; (6) aldrin; (7) heptachloroepoxide; (8) 4,4'-DDMU; (9) dieldrin; (10) 4,4'-DDE; (11) endrin; (12) 4,4'-DDD; (13) 2,4'-DDT; (14) 4,4'-DDT; (15) mirex. Remaining peaks in (A) are PCB

bornene have been identified as contaminants in fish from the Mississippi River[153] and the Wabash River.[154]

The multiresidue determination of volatile and semivolatile organochlorine compounds in soil and chemical waste disposal site samples with a lower limit of detection of 10 μg/g was studied.[155] The procedure involves a simple hexane extraction followed by HRGC WCOT column coated with SE-52 (25 m $\times$ 0.3-mm i.d.) using ECD and GC/MS. A comprehensive analysis of halogenated compounds in many and the environmental media, in the breath, blood and urine of an exposed population in the Love Canal are in Niagara Falls, New York, was extensively studied.[156] In addition, levels of halogenated hydrocarbons in air samples taken in Buffalo and Niagara Falls were determined using a 100-m long SE-30 SCOT column. Residue analysis of organochlorine hydrocarbons in human bone marrow containing hexachlorocyclohexane isomers was performed by means of HRGC with ECD. The WCOT columns employed were 30 m $\times$ 0.3-mm i.d. coated with SE-30 and Apiezon L.[157] Chromatograms are shown in Fig. 8.16.

A method for the determination of very low concentrations (pg/m^3) of chlorinated hydrocarbons in air samples from the Arctic using polyurethane foam as a collection medium was devised.[158] For quantitation and identification, negative ion CIMS and ECD were combined with a WCOT column (30 m $\times$ 0.3-mm i.d.) coated with SE-30, SE-54, and OV-17.

8.6.3. Analysis for Mirex and Kepone

The discovery of mirex (dodecachloro-octahydro-1,3,4-methano-2H-cyclobuta-[c,d]pentalene) in Lake Ontario fish sparked considerable effort to monitor mirex and its degradation products in sediment and fish.[159] The development of a nitration method improved the ability to detect much smaller quantities of mirex and its degradation products than was possible by previous methods.[160] Nitrated extracts of herring gull eggs were analyzed by WCOT column GC and GC/MS. Samples were injected on a 20-m OV-17 WCOT column and run at 180 °C isothermally. Five minor mirex-related compounds were identified in addition to mirex and the previously identified 8-monohydromirex.[161] Quantiative determination of mirex and its degradation products by WCOT column GC/MS in SIM mode under electron impact ionization in lake trout samples from Lake Ontario and in lamprey samples and suspended sediment samples was also reported. The comparison of results obtained by SIM and ECD shows insignificant differences between results.[162] SIM traces for an injected standard of 100 pg of each mirex degradation product are shown in Fig. 8.17.

Mirex is a caged dimer of a hexachlorocyclopentadiene and has been

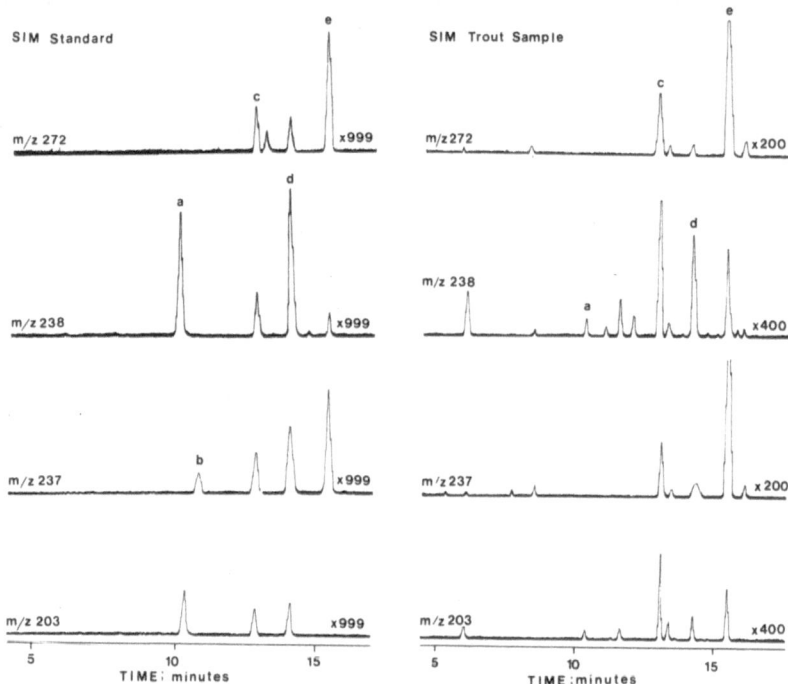

FIGURE 8.17. SIM traces of mirex and its degradation products from fish samples. (a) 2,8-dihydro; (b) 10,10-dihydro; (c) 8-monohydro, (d) 10-monohydro; (e) mirex.

used as both flame retardant and insecticide. Photomirex, the 8-monohydro- and 2,8-dihydromirex are formed by photolysis or reductive dechlorination of mirex.[163] Mirex analysis is often complicated by the presence of PCB and other organochlorine insecticides. Even GC/MS determination of mirex presents some difficulties; the major fragment, corresponding to $C_5Cl_6^+$, can arise from a variety of compounds, particularly from PCB isomers, and care in the cleanup procedure must be exercised.[164]

Kepone (1,1a,3,3a,4,5,5,5a,5b,6-decachloro,octahydro-1, 4-methano-2H-cyclobuta[c,d]pentalen-2-one), the caged 2-keto analog of mirex, has been extensively used in pesticide formulations, and a severe contamination has resulted. Gas chromatography–mass spectrometry involving chemical ionization and high-resolution mass spectrometry were employed to detect, identify, and confirm the presence of Kepone.[165] Kepone and its degradation products could be analyzed on a OV-101 WCOT column together with various pesticides as shown in Fig. 8.15.[166] The choice of solvents is critical in the extraction step, cleanup, and analysis of Kepone. Benzene, acetone, and methanol are better solvents than aliphatic hydrocarbons. However, methanol apparently reacts with Kepone to form Kepone hemiacetal, and it is possible that this could attribute to the increase of electron affinity in the ECD. An analytical procedure for the determination of Kepone in water and sediment is described.[167] The lower limit of detection in water is 20 ng/l and 10 μg/kg in sediments. Average recovery of Kepone from samples spiked with 4 mg/kg is 103%.

8.7. HALOGENATED AROMATIC HYDROCARBONS

Highly volatile organochlorine compounds such as chlorinated benzenes have been recognized as persistent contaminants at higher trophic levels in large aquatic ecosystems all over the world. These compounds have high octanol–water partition coefficients, and it can be expected that they would be bioaccumulated in the aquatic ecosystem.[168] Many of them have been used as solvents, intermediates for organic synthesis, herbicides, seed disinfectants, and additives for PCB-filled transformers.

Gas chromatography has played a major role in the study of the chemistry, metabolism, breakdown, and environmental occurrence of chlorinated aromatic industrial contaminants. Halogenated aromatic hydrocarbon pesticides and their metabolites were the first group of organic compounds restricted due to their persistence and bioaccumulation in our environment, when analyses for DDT and related compounds has confirmed their ubiquity in the ecosystem.

8.7.1. Chlorinated Benzenes

Chlorinated benzenes are prevalent in industrial effluents,[168] atmospheric discharges, tributaries, sediments, and wildlife.[169] They have been found in the adipose tissue of pigeons and humans in Japan,[170] and in the fly ash and flue gases of municipal incinerators in Japan and Canada.[171] Lake Ontario herring gull eggs and tissues, coho salmon, smelt, and alewives have consistently shown evidence of chlorinated benzenes. Hexachlorobenzene (HCB) has been determined at levels greater than 5 μg/ kg in fish.[154,172] Pentachlorobenzene was reported in North Sea fish[173] and fish from Norwegian fjords.[174] All isomers were identified in water,[10] sediment,[168,175] herring gulls,[169] and in some seeds.[176]

The determination of chlorinated benzenes is accomplished with various WCOT columns such as SP-2100, SE-30, OV-17, and Carbowax 20M using ECD,[168,177] and GC/MS using negative ion CIMS.[178] Relative retention times for four different WCOT columns are given in Table 8.19. Relative retention times of some brominated benzenes on SE-52 WCOT column are given in Table 8.20.

TABLE 8.19
Relative Retention Times of Chlorinated Benzenes[a]

Isomers	SP-2100	DB-1	SE-52	CX-20M
Chlorobenzene	0.273		0.128	0.294
1,3-dichlorobenzene	0.437	0.204	0.181	0.465
1,4-dichlorobenzene	0.446	0.207	0.177	0.490
1,2-dichlorobenzene	0.469		0.196	0.519
1,3,5-trichlorobenzene	0.562	0.303	0.278	0.544
1,2,4-trichlorobenzene	0.603	0.349	0.326	0.637
1,2,3-trichlorobenzene	0.634	0.385	0.366	0.696
1,2,3,5-tetrachlorobenzene	0.732	0.531	0.506	0.734
1,2,4,5-tetrachlorobenzene	0.733	0.533	0.506	0.741
1,2,3,4-tetrachlorobenzene	0.768	0.589	0.571	0.819
Pentachlorobenzene	0.870	0.782	0.755	0.882
Hexachlorobenzene	1.000	1.000	1.000	1.000

[a] Conditions: SP-2100; 30 m in length, 0.25-mm i.d., ($d_f = 0.3$ μm), nitrogen as a carrier gas at 1.3 ml/min; 33 °C initial, hold 3 min; TP 10–180 °C; DB-1; 30 m × 0.25-mm i.d. ($d_f = 0.1$ μm), hydrogen as a carrier gas at 1 ml/min; 75 °C initial, hold 2 min; TP 4 °C/min to 170 °C; SE-52; 15 m × 0.25-mm i.d. carrier gas methane; TP 60–240 °C at 4 °C/min at $u_o = 25$ cm/sec; C-20 M; 30 m × 0.25-mm i.d. ($d_f = 0.08$ μm) nitrogen as a carrier gas; 33 °C initial, hold 2 min; TP 10–180 °C.

Methodology for the determination of chlorinated benzenes in sediments by means of the steam distillation cleanup method and WCOT column was developed.[179] The sediment sample is extracted by means of steam distillation from organics–free water added to the sediment and is partitioned with n-hexane in a continuous Nielson–Kryger steam distillation apparatus with an attaching disposable pippette containing a Tenax trap. The reducing adaptor and Tenax trap is set on the top of the Nielson–Kryger condensor to trap lower molecular weight chlorinated benzenes. The combined extracts (n-hexane and 2 ml of hexane passed through the Tenax trap) are dried, reduced in volume, and analyzed by HRGC–ECD (see Fig. 8.18).

TABLE 8.20
Brominated Benzenes Relative Retention Times on SE-52 WCOT Column

Compound	RRT[a]
Bromobenzene	0.140
1,4-dibromobenzene	0.334
1,3-dibromobenzene	0.335
1,2-dibromobenzene	0.363
1,3,5-tribromobenzene	0.625
Tetrabromobenzene	1.028
Hexabromobenzene	1.780
Hexachlorobenzene	1.000

[a] GC conditions the same as in Table 8.19.

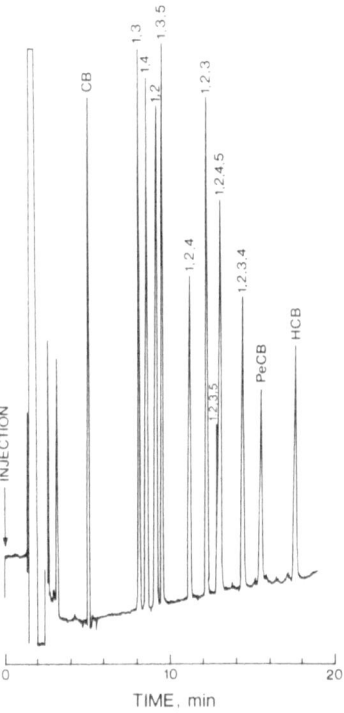

FIGURE 8.18. Chlorobenzene mixture analyzed on Carbowax 20M WCOT column. Peaks and concentrations in μg/l: (CB) chlorobenzene: 30,000; (1,3) dichlorobenzene: 74; (1,4) dichlorobenzene: 132; (1,2) dichlorobenzene: 72; (1,3,5) trichlorobenzene: 13; (1,2,4) trichlorobenzene: 8.1; (1,2,3) trichlorobenzene: 7.4; (1,2,3,5) tetrachlorobenzene: 1.3; (1,2,4,5)-tetrachlorobenzene: 8.9; (1,2,3,4) tetrachlorobenzene: 2.9; (PCB) pentachlorobenzene: 1.4; (HCB) hexachlorobenzene: 1.2. Splitless injection; initial temperature 33 °C; hold for 3 min; temperature programmed at 10 °C/min to 180 °C; SP-2100 WCOT column 30-m × 0.25-mm i.d. (d_f = 0.08 m). (Chromatogram courtesy of Dr. B.G. Oliver, National Water Research Institute, Burlington, Ontario.)

Studies of the global baseline pollution of the sub Antartic area on penguins as bioindicators of the marine environment was investigated by Ballschmiter.[180] The different species such as the Gentoo penguin (*Pygoscelis papua*), Rockhopper penguin (*Eudyptes crestatus*), and the Magellanic penguin (*Spheniscus magellanicus*) and their eggs from the Falkland Islands were found to be contaminated with α- and δ-hexachlorocylohexane, hexachlorobenzene, PCBs, 2,4'-DDT, 4,4'-DDT, 4,4'DDE, 4,4'-DDD, mirex, and polychlorinatd terpenes. In eggs of the black-browed albatross (*Diomedea melanophris*), technical chlordane constituents were identified, and in eggs of the Kelp goose (*Chloephaga hybridea*) traces of HCB and 4,4'-DDE were detected. A fused-silica WCOT column 25-m in length coated with SP-2100 was used. The components of the sample were identified by comparison of their retention times with those of standard mixtures.

Organochlorine compounds residues containing HCB, α-, β-, γ-, and δ-hexachlorocylohexane (HCH) besides other chlorinated pesticides were separated on the OV-17 WCOT column (20 m × 0.25-mm i.d.) using temperature programming from 80 to 200 °C at 7 °C/min and ECD.[148]

A multicomponent mixture of polychlorinated styrenes consisting of hexachloro-, heptachloro- and octachlorostyrenes were identified in fish from both European and North American waters. These compounds are not commercially produced but are formed as side-reaction products during high-temperature carbon–chlorine reactions.[147] These contaminants have been found in fish from Lake Ontario, Lake Huron, and the Detroit River.[172,181] The contamination of the Norwegian fjords resulted from the industrial process converting magnesium oxide to magnesium chloride.[182]

There is no question that 1,1,1-trichloro-2,2-bis-(p-chlorophenyl) ethane (DDT) and its analogs have been the most recognized contaminants in the global ecosystem. Chemical substances shown below deposited in the biosphere and troposphere are transported often thousands of kilometers from their original site of application.

The chemical formula of DDT isomers and related compounds are shown in Scheme 8.1.

Their rates of utilization were such that these compounds contaminated water, soil, and food in concentrations of μg/kg to mg/kg, depending on the location and the meteorological conditions. Most of these contaminants were concentrated in organisms and in lipids at each trophic level. Low detection limits are often demanded for their determination. This can make it difficult to discriminate between the insecticides of interest and other contaminants on the chromatogram. The use of WCOT columns for analyzing the majority of these contaminants is superior as shown in Fig. 8.16B.

The solubility of DDT isomers and its analogs in water is very low. These compounds can be extracted with water-immiscible solvents such as benzene or hexane. Usually several hundred liters of water must be extracted. Recoveries at the 0.2–340 μg/l level are between 83 to 100%.[183] Several applications of WCOT columns to the analyses and determination of halogenated aromatic hydrocarbons in drinking water,[184] surface waters,[184,185] sediment and soil,[185] sewage,[186] and human and animal tissues[187,188] have been described.

Halogenated aromatic hydrocarbons were analyzed, employing the ECD or GC/MS in selected ion monitoring mode to attain high sensitivity. These compounds can easily be detected in the low μg/kg range.

Many types of stationary phases are used to separate the chlorinated aromatic hydrocarbons. An OV-17 is an excellent liquid phase for their analysis because of its high temperature stability and medium polarity, thus permitting resolution of many organochlorinated (OC) pesticide hydrocarbons. It is sometimes advisable to employ higher polarity coatings in order to resolve multicomponent spray mixtures. Since the newer

SCHEME 8.1

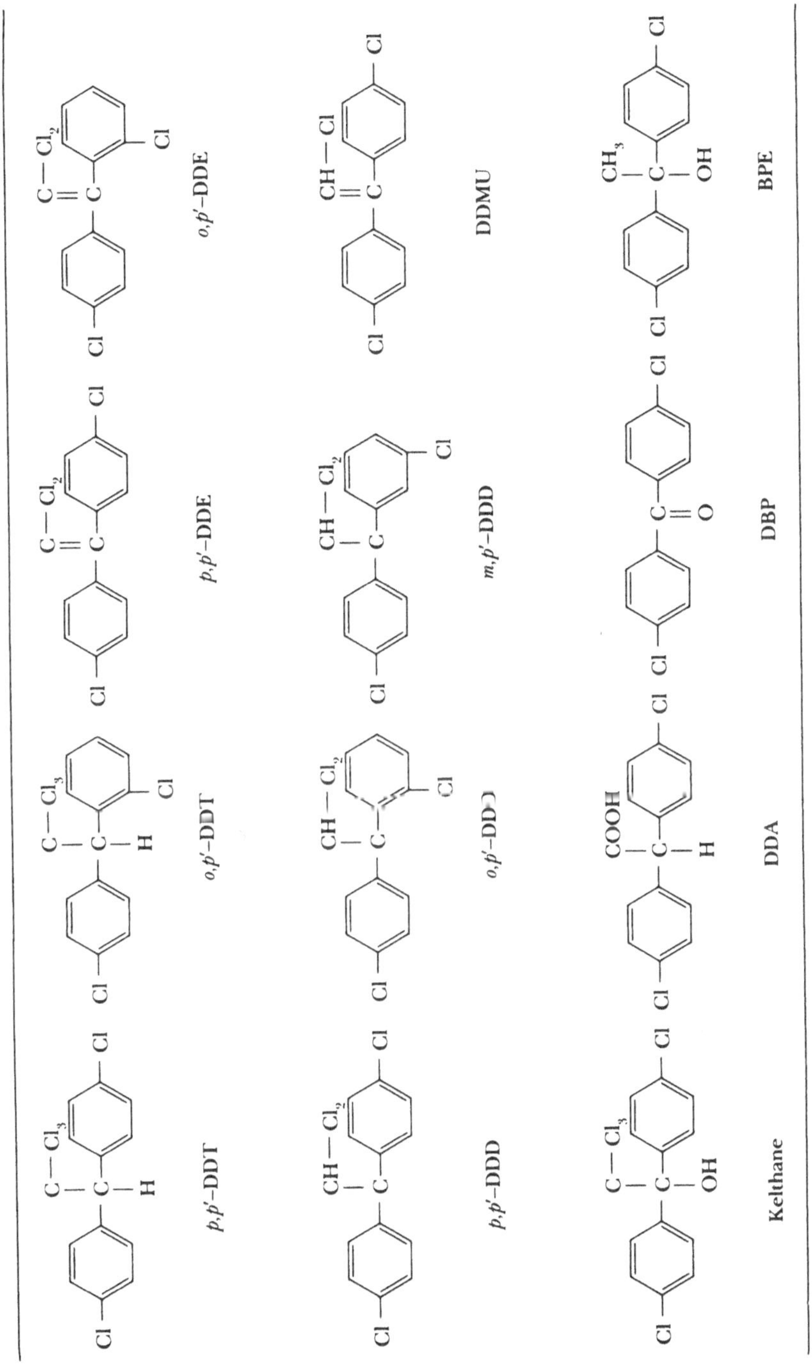

gum phases of OV-210 type are available commercially on fused-silica columns, they can be used alone or in mixtures with OV-17 and OV-1.

8.7.2. Polychlorinated and Polybrominated Biphenyls and Terphenyls

Analysis of polychlorinated biphenyls (PCB) as environmental contaminants is complicated by the presence of a large number of congeners in the commercial mixtures and their noncommercial formulations. Significant differences exist in the toxicity of individual PCB isomers. To date, WCOT column gas chromatography has provided the most complete separation of PCB congeners in technical mixtures and environmental samples.

Particulate matter from air samples can be trapped on a suitable filter, such as polyurethane foam[158] or nylon mesh (200 μm) coated with a silicone oil, and the PCB are then re-extracted with n-hexane.[189] Extraction schemes are shown in Chapter 5 for water and biological samples. Most analytical chemists have used standard methods for OC in which the PCB are extracted together with fat. The sample is grounded with anhydrous Na_2SO_4 and extracted with n-hexane.

Cleanup methods for removal of fat from the extract include solvent partitioning between n-hexane and acetonitrile, or saponification of fatty tissues with ethanolic potassium hydroxide. Gel permeation chromatography has also been employed.[190] PCB may be separated from OC pesticides by column chromatography on Florisil. A microcleanup-simplified technique has also been proposed.[191]

For purpose of further discussion terms such as isomer, homologue, and congener will be defined. They are used not only with PCB mixtures but also with chlorinated dibenzo-p-dioxins, chlorinated dibenzofurans, halogenated diphenyl ethers, chlorinated naphthalenes, and chlorinated phenols.

An isomer is defined by means of the numerical arrangement of a substituent, such as halogen atoms, within the moiety of the homologue, e.g., 2,5,2′,5′-tetrachlorobiphenyl isomer. The homologue represents a group of isomers having a specific number of halogen atoms; thus the tetrabromobiphenyl homologue series has 30 possible isomers. The congener as a term will mean any isomer of any homologue; thus, there are a total of the 209 possible PCB congeners.

8.7.2a. Qualitative PCB Analysis. The high resolution of WCOT columns can be employed as a helpful tool for identification of single components in complex mixtures. Matching of Kovats retention indices at isothermal conditions or relative retention times (RRT), when temperature-programming conditions and ECD detection is employed, is a well-established mean for qualitative identification when rigorously constant conditions are used.[192-194] Besides the constancy of the GC parameters, care must be taken for an accurate determination of net retention times when Kovats retention indices are measured. Of course, a problem exists of how to measure a dead volume or void retention time (t_a) when ECD is employed. It is known that the ECD does not respond to hydrocarbons essential for the measurement. In this case the relative retention times can be measured. The reproducibility of RI determination strongly depends on the polarity and on the adsorptivity of the WCOT column. Provided that a sufficiently deactivated and conditioned WCOT column is available, an accuracy of retention time measurement of better than 0.05% retention index can be achieved.[195] These authors also performed a conversion of the n-alkanes based RI concept to the use of the series of n-alkyl-trichloroacetates (ATA), which allows use of the ECD in the picogram range. Retention indices of PCB isomers on Apiezon L coated WCOT column 40 m $\times$ 0.3-mm i.d. with a β value of 1000 at 180 °C are given in Table 8.21.[196]

The most common stationary phases used for PCB analysis are Apiezon L,[197] Apiezon M,[198] Kovats C-87 hydrocarbon,[199] OV-101,[192,193,200] OV-17,[193] and SE-54.[201]

8.7.2b. Quantitative Analysis of PCB Residues. The response of ECD is not the same for all PCB congeners but is much affected by the degree of chlorination on the biphenyl moiety.[202] As a result, it is important to consider not only the total PCB residue level in each sample but also the quantitative distribution of the various congeners present in the different PCB homologues. This is necessary because once PCB is released into the environment the original congeners distribution pattern of any formulation is altered as a result of specific changes in the environmental conditions. Often, complications are observed in determining the PCB residue profiles when different Aroclor mixtures are present in a sample. Difficulties are encountered when the PCB in the matrix have undergone selective degradation. It was observed that the pattern of chromatographic peaks from such samples is similar to the one resembling the higher chlorinated Alochlor mixtures. Thus, these samples are determined and compared with those of the most resembling commercial product or their mixtures to obtain the amount of PCB in the sample.

Quantitative determination of PCB residues can be performed by the following techniques using either ECD, the Hall detector, or GC/MS–CI techniques:

(i) using hydrodechlorination of PCB congeners to biphenyl by use of $(Ni_2B)_2H_3$,[203]

TABLE 8.21
Kovats Retention Indices of PCB Iosmers[a]

Iosmer	I	Isomer	I
α-HCH	1715.9	Endrin (cont.)	
HCB	1749.5	22'355'6	2192.0
γ-HCH	1770.2	22'34'56	2200.7
22'3 + 24'6	1830.0	22'33'56'	2203.7
2'35	1864.0	22'345'6	2211.7
24'5 + 244'	1882.1	22'34'5'6	2215.5
233'	1898.3	4,4'DDD	2222.7
22'46'	1909.1	22'33'5'6'	2226.1
Heptachlor	1913.2	23'44'5	2230.4
33'5	1929.6	24'DDT	2233.8
22'55'	1945.2	22'33'46	2233.9
23'46 + 22'45'	1952.8	22'3456'	
22'44'	1958.5	22'33'55'	2241.2
Aldrin	1975.1	22'33'46'	2252.6
22'35'	1977.7	233'44'	
344' + 22'34'	1983.6	22'44'55'	2263.9
234'6	2000.4	22'33'566'	2282.3
22'33'	2011.0	22'3455'	
23'45	2032.1	22'33'45'	2294.0
Heptachlorepoxid	2042.1	22'344'5	
244'5 + 233'5'	2046.0	4,4'-DDT	2297.3
23'4'5	2051.3	22'344'5'	2304.7
22'35'6 + 23'44'	2056.5	22'33'45	2316.0
22'34'6	2064.7	22'33'55'6	2329.6
22'456'	2067.8	22'33'44'	2344.7
2344'	2082.5	22'34'55'6	2353.9
22'33'6 + 22'44'66' +	2089.6	233'455'	2364.7
22'35'		22'3455'6	2369.4
22'34'5	2094.6	22'344'5'6	
22'455'	2106.0	22'33'4'56	2386.8
22'44'5	2113.9	233'44'5	2395.6
22'3'466'	2125.6	22'33'456	2397.6
22'3566'	2128.5	22'33'55'66'	2404.7
22'3'45	2138.4	22'33'45'66'	2416.2
233'5'6		22'33'44'66'	2427.1
Dieldrin	2144.5	22'33'455'	
22'345'	2145.2	22'344'55'	2439.2
22'344'	2149.7	22'33'44'5	2481.1
22'33'66'	2152.0	22'33'455'6	2510.3
4,4'-DDE	2156.2	22'33'455'6'	2520.2
2,4'-DDD	2162.3	22'33'44'56'	2544.7
33'44'	2163.3	22'33'44'56	2563.7
Endrin	2174.4	22'33'44'55'	2612.1
22'33'4			
23'45'5	2177.6		
233'4'6			

[a] Recalculated on n-alkyl-trichloro acetates basis, where n-hexyl-trichloroacetate is equal to 1400. Conditions: 40 m × 0.3-mm i.d. SE-30; SN = 45 for C_{12}–C_{18} pair at 70 °C; column temperature at 30 °C injection, within 2 min to 15 °C, 3 min at 150 °C and programmed at 1 °C/min to 200 °C; injector temperature 250 °C; detector: 270 °C.

(ii)   perchlorination of PCB congeners to decachlorobiphenyl,[204,205]
(iii)   total PCB using multiple internal standards,[198,201]
(iv)   quantitation of homologues,[194] and
(v)   determination of individual isomers.[201,206,207]

The first two methods assume a complete conversion to only one product and serve as a total PCB level value for the total loading data. The third methodology also provides the total PCB content, but it may be used for determining a few individual congeners also. The fourth one is capable of providing all three answers, not only the total PCB residue in each sample but also to group the GC peaks in their homologue sets according to the number of chlorine atoms in the biphenyl moiety as determined from the previous GC/MS data. The PCB homologue contents are calculated from the ECD responses to the 22 selected isomers representing approximately 50% of the PCB concentration in the sample. The remaining 50% are averaged out as total homologue relative response factors to the known concentration of a PCB mixture (e.g., 300 pg/μl of 1:1:1 Aroclor 1242, 1254, and 1260) (Fig. 8.19). This method can provide also quantitation of individual congeners. The method requires cold on-column injection, and decachlorobiphenyl is employed as a marker for determining all RRTs relative to this isomer. A computing integrator, such as SP-4100, has built-in programs capable of providing reliable results.

A different methodology in which isomer-specific PCB determination offers a means of comparing WCOT column GC records by the computer program PEAK-11 in which files of reviewed, edited, and quantitated data are treated by multivariable statistics has been recommended.[206,207] A component pattern recognition technique (SIMCA) was applied to the problem of establishing similarities among, and differences between, environmental PCB profiles. Relations among samples can be plotted in two-dimensional data projection, which allows comparison of similarities among groups of samples having the same or different residue patterns.

Quantitative determination of PCB residues represents quite a difficult task, and it must be realized that there is no HRGC methodology that has been properly validated. For this reason caution should be taken in accepting the analytical results from laboratories until their competence has been established and the laboratory is certified for this type of multiresidue analysis.

Confirmation of PCB residues can be performed by GC/MS—SIM using the chemical ionization mode of operation[208] (Fig. 8.20). The results show that such an approach can be successfully adapted to identify and quantitate PCB mixtures. Representative congeners from each homologue series can be utilized as single standard to quantitate all congeners

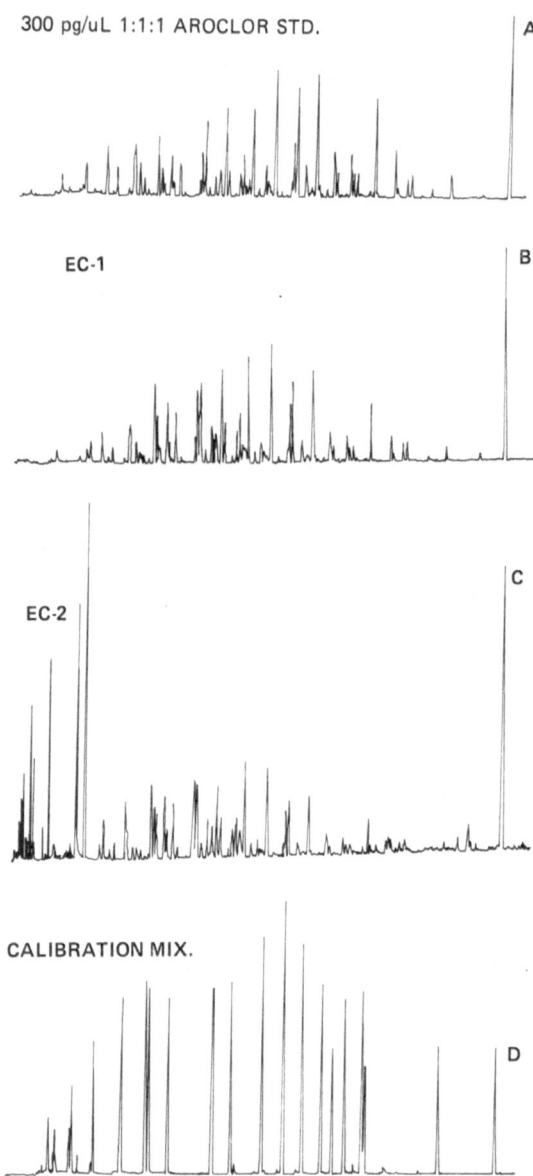

FIGURE 8.19. PCB analysis showing (A) 300-g Aroclor 1242, 1254, 1260 (1:1:1) mixture; (B) sediment sample extract EC-1; (C) sediment sample extract EC-2; and (D) calibration mixture of 22 PCB isomers. WCOT-FS cross-linked SE-52 column 60 m × 0.32-mm i.d. (d_f = 0.25 μm); hydrogen carrier gas; temperature programmed from 75 °C (hold for 2 min), ballistically to 140 °C and programmed to 240 °C at 2 °C/min; cold on-column injection; ECD.

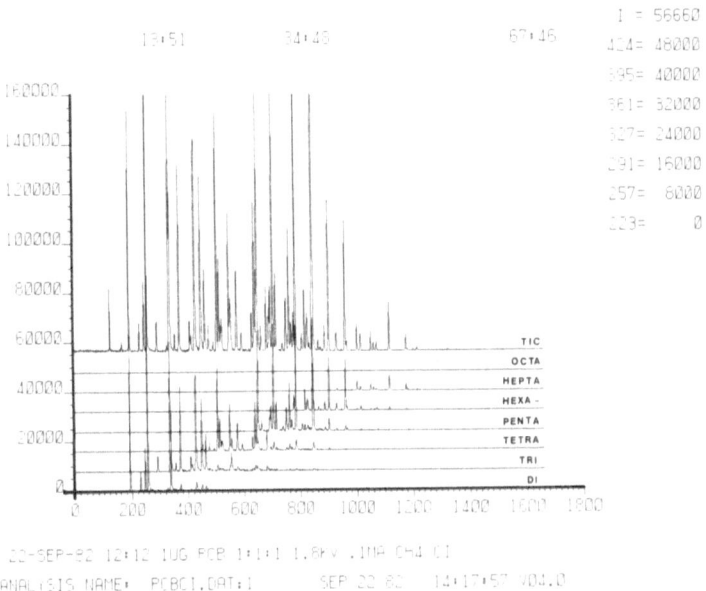

FIGURE 8.20. Mass chromatogram of PCB Aroclor 1242, 1254, 1260 (1:1:1) mixture obtained by GC/MS–CI for identifying the number of chlorine substituents on the biphenyl moiety.

of their respective homologues. The MH⁺ ions corresponding to the most abundant ions in their CI mass spectra are:

Monochlorobiphenyls	m/z	189; 191
Dichlorobiphenyls	m/z	223; 225
Trichlorobiphenyls	m/z	257; 259
Tetrachlorobiphenyls	m/z	293; 291
Pentachlorobiphenyls	m/z	327; 325
Hexachlorobiphenyls	m/z	361; 359
Heptachlorobiphenyls	m/z	395; 393
Octachlorobiphenyls	m/z	429; 427
Nonachlorobiphenyls	m/z	463; 461

These ions can be used for SIM analysis by GC/MS. Electron impact SIM GC/MS can also be used for the homologues and determination of single PCB isomers. The results show small variability in response within the homologue series, and the average response factors were proposed as a basis for their quantitative determination.[209]

8.7.3. Polybrominated Biphenyls

Polybrominated biphenyls (PBB) have received relatively little attention, even though they have been identified in the state of Michigan where they contaminated animal feed. This contamination resulted in the destruction of many animals and a commercial formulation of fireMaster BP-6 consisting mainly of 2,4,5,2',4',5'-hexabromobiphenyl was positively identified.[210] Also in New York, New Jersey, and in fish from the Ohio River some contamination with lower brominated biphenyls and decabromobiphenyl was found.[21]

Since there are also 209 possible PBB congeners in technical products, one would expect to face the same separation problems as with PCB mixtures. Until now, separation of fireMaster BP-6 indicates that a WCOT column 10 m in length coated with OV-101 resolves very well all components in the sample, except that 2-hydroxybiphenyl peak is strongly retained on the active column used in this case.

8.7.4. Polychlorinated Terphenyls

Polychlorinated terphenyls (PCT) should always be present together with PCB as side-reaction products. It is known that technical biphenyl used for chlorination contains between 6 to 15% of o-, m-, and p-terphenyls and higher polyphenyls.[212] These three isomers of terphenyl are always present in this mixture. It is obvious that these impurities undergo the same reaction mechanism as biphenyl. On the other hand these compounds were also manufactured under various trade names such as Clophenharz in Germany, Aroclor 5460 in the United States, and Delorene in Czechoslovakia.

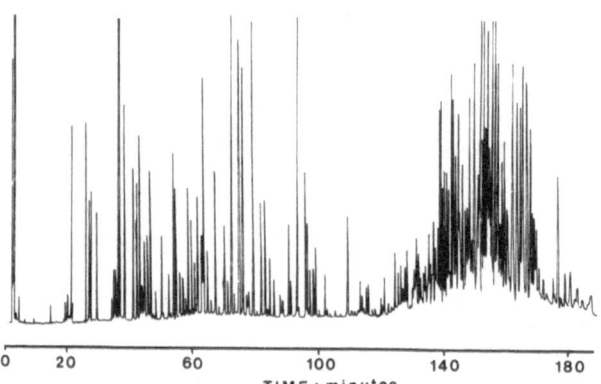

FIGURE 8.21. Mixture of chlorinated terphenyls (Arocolor 5460) and PCB (Aroclors 1016, 1254, 1260) in 10:2.5:2:1 ratio. WCOT-FS column deactivated with SE-54 and coated with Apiezon M ($d_f = 0.025$ μm). Temperature programmed from 100 to 270 °C at 1 °C/min, hold for 20 min. Carrier gas helium at 40 cm/sec.

Identification of PCB at trace levels in human adipose tissue in autopsy and surgical samples require a thorough cleanup. Gel permeation chromatography on the 35-cm × 25-mm i.d. column containing BioBeads SX-3 (BioRad Laboratories, Richmond, California) was employed using toluene in ethyl acetate (1:3 v/v) at a flow rate of 5 ml/min. The first 100 ml is discarded. The subsequent fraction (90 ml) contained pesticides residue and PCT.[213]

The complexity of the chromatogram shown in Fig. 8.21 indicates difficulties of quantitation and confirmation since the number of possible congeners would be enormous. There are 14 hydrogen atoms replaced by chlorine atoms versus 10 hydrogen atoms available in biphenyl.

8.8. POLYCHLORINATED DIBENZO-P-DIOXINS

Polychlorinated dibenzo-p-dioxins (PCDD) are ubiquitous impurities identified in terrestrial and aquatic environments. There are 75 possible PCDD congeners, some of which are known to be very toxic, especially the 2,3,7,8-tetrachlorodibenzo-p-dioxin (2,3,7,8-TCDD) isomer with the structural formula of:

This molecule is symmetrical across both vertical and horizontal axis and its chemical stability is remarkable. Because of the different substitution, the homologue series and corresponding congeners exist as shown in Table 8.22.

Analytical methods for detecting PCDD in various types of samples involve extensive sample preparation procedures followed by highly complex instrumental analysis.[214] Ultratrace analytical determination of individual isomers of PCDD requires specific extraction and cleanup techniques, which are still the subject of validation and controversies among researchers.

8.8.1. Sample Preparation

In order to determine PCDD in various matrices at low levels (μg/kg and less), these matrices must be removed and purified. The sample preparation procedures may be divided into the following steps:

1. Initial sample treatments that include:
   a. direct extraction[215–223]

TABLE 8.22
Molecular Formula, Molecular Weight, and Distribution of Iosmers of Polychlorinated Dibenzo-p-Dioxins

Homologue	Molecular formula	Molecular weight	Number of isomers
Monochloro	$C_{12}H_7ClO_2$	218.0133	2
Dichloro	$C_{12}H_6Cl_2O_2$	251.9744	10
Trichloro	$C_{12}H_5Cl_3O_2$	286.2865	14
Tetrachloro	$C_{12}H_4Cl_4O_2$	319.8965	22
Pentachloro	$C_{12}H_3Cl_5O_2$	353.8577	14
Hexachloro	$C_{12}H_2Cl_6O_2$	387.9592	10
Heptachloro	$C_{12}HCl_7O_2$	421.7799	2
Octachloro	$C_{12}Cl_nO$	455.7410	1

    b.   saponification in acids[224] and in base[221,225]

    c.   dissolution[219,226–230]

2. The liquid–liquid partitioning includes:

    a.   partitioning against aqueous base[216,219,226,228,230,231]

    b.   partitioning against conc. H_2SO_4.[221,222,230,232,233]

3. Column chromatography is performed on substrates as:

    a.   alumina[215–217,221,222,224–228,231,234–236]

    b.   silica gel[216,219,221,222,228,229,236]

    c.   Florisil[215,233,235]

    d.   ion exchangers[219,230]

    e.   charcoal and active carbon[220,237,238]

    f.   gel permeation[223,236–238]

    g.   reagent modified silica.[216,224,225]

4. High-pressure liquid chromatography for trapping the appropriate retention volume containing PCDD only.[224,225,228,233]

5. Gas chromatography has been the final separation tool and both packed and WCOT columns have been employed but the WCOT columns provide more isomer specificity.[239–241]

A study to evaluate extraction and cleanup procedures for determining 2,3,7,8-TCDD in fish has been carried out.[242] It was concluded that two procedures[225,236] provide efficient cleanup since no interferences and low levels of coextractives were observed. These two procedures allow screening samples by ECD WCOT column GC.

8.8.2. Qualitative PCDD Analysis

For trace level PCDD analyses, qualitative and quantitative accuracy and precision is highly dependent upon chromatographic performance. An ideal WCOT column for separating PCDD congeners would have no activity and would provide adequate resolution with minimum analysis time. Major emphasis in analyzing environmental samples has been on determining the 2,3,7,8-TCDD isomer. Unfortunately, there are at least three different congeners eluting with very close retention time. We have already shown that while the relative retention (α) depends only on the liquid phase and temperature, the resolution (R) between two adjacent peaks is dependent on the WCOT column efficiency. Utilizing relationships described in Chapter 1, we can also extend them to calculate the required column length (L_{req}) if the HETP ($=L/n$) is known.

$$L_{req}/H = n_{req} = 16 \cdot R^2 \, (\alpha/\alpha - 1)^2 \, (k' + 1/k')^2$$

where H, n, and k' refer to the second peak.

In the early work of Buser[231], three different polarity stationary phases, OV-101, OV-17, and Silar 10c, were studied. Thin film coatings ($d_f = 0.1$ μm) but relatively wide-bore tubings were employed. Even so, greatly increased separation efficiencies were achieved on WCOT columns having on the average of less than 2000 plates/m calculated for octachlorodibenzo-p-dioxin at 225 °C. Electron capture and MS–SIM detectors were used for analyses. Elution patterns are shown in Fig. 8.22, and elution temperatures including relative retention data are given in Table 8.23.

Silar 10C allows the highest number of congeners to be qualitatively separated from the most toxic 2,3,7,8-TCDD. However, at present, there is no one WCOT column tested to resolve all 22 possible isomers, even though the technology, theoretical knowledge, and skills are available to produce a WCOT column having 350,000 TP for 2,3,7,8-TCDD.

The relative retention times for 39 PCDD using the 50-m $\times$ 0.2-mm-i.d. SP-100 WCOT column at 225 °C were reported. Helium was employed as a carrier gas at 0.5 ml/min.[246] Carbowax 20M seems to be also a selective phase to separate PCDD isomers.[241] The elution pattern of all TCDD isomers is shown in Fig. 8.23. Qualitative TCDD isomer assignments capable of separating TCDD isomers on three different WCOT columns are given in Table 8.24.

Unfortunately, the qualitative information in this study does not give pertinent data such as Kovats retention indices or RRT values with elution temperatures.[231] For this reason an analyst must rely on his own set of standards and use these data as a guide.

8.8.3. Quantitative PCDD Analysis

The majority of PCDD data collected are obtained using HRGC/MS–SIM. These data cover combustion processes where point-source

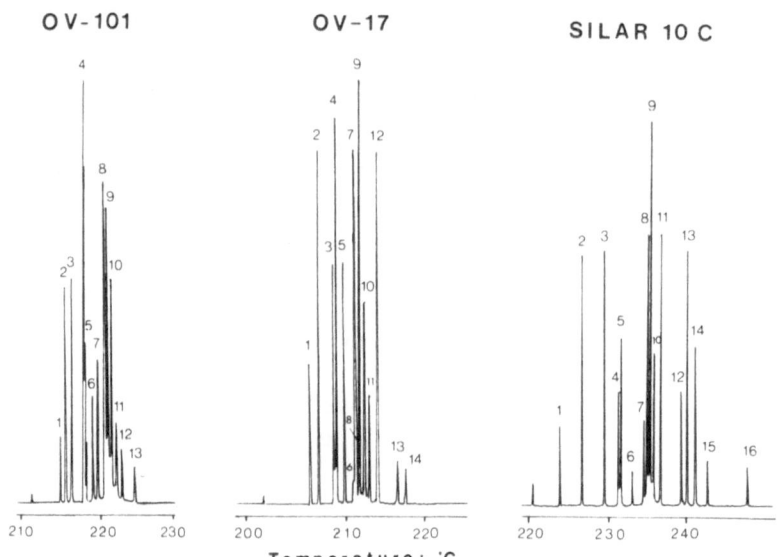

FIGURE 8.22. Separation of polychlorinated dibenzo-*p*-dioxins and dibenzofurans on three different WCOT columns (A) 55 m × 0.36-mm i.d. OV-101; (B) 55 m × 0.32-mm i.d. OV-17; and (C) 55 m × 0.35-mm i.d. *Silar-10 C*: (1) 1368; (2) 1379; (3) 1378; (4) 1369; (5) 1247 + 1248; (6) 1268; (7) 1478; (8) 2378; (9) 1234 + 1246 + 1249; (10) 1237 + 1238; (11) 1236 + 1278; (12) 1278 + 1469; (13) 1239; (14) 1269; (15) 1267; (16) 1289. *OV-17*: (1) 1368; (2) 1379; (3) 1369; (4) 1378; (5) 1247 + 1248; (6) 1268; (7) 1246 + 1249 + 1478; (8) 1469; (9) 2378 + 1279; (10) 1237 + 1238 + 1234; (11) 1236; (12) 1269 + 1239; (13) 1278; (14) 1267; (15) 1289. *OV-101*: (1) 1368; (2) 1379; (3) 1369; (4) 1378 + 1469 + 1247 + 1248; (5) 1246 + 1249 + 1268; (6) 1478; (7) 1279; (8) 1234 + 1236 + 1269 + 1237 + 1238; (9) 2378; (10) 1239; (11) 1278; (12) 1267; (13) 1289.

emissions and ambient levels are studied.[243] Basically, samples associated with emissions from incinerators include slag, quench, fly ash, water, flue gas particulates, and gaseous emissions.[244] The PCDD content of human milk and fish samples have also been reported.[235]

Low- and high-resolution mass spectrometry have been utilized, and both the quadrupole and magnetic sector instruments, single- or double-focusing systems, have been employed. The use of electron impact ionization mode prevails, although recently the positive and the negative ion CIMS techniques give good results.[223,240,245]

8.8.4. Application of WCOT Columns in PCDD Analysis

Methods for determination of PCDD in various types of samples have appeared in the literature. Selected references illustrate applications of

TABLE 8.23
PCDD Relative Retention Times on SP-2100 WCOT Column and Elution Temperature
on OV-17 WCOT Column

Isomer	Substitution pattern	RRT SP-2100	Elution temperature (°C) OV-17
1-MCDD	1:0	0.293	
2-MCDD	1:0	0.299	169.3
2,7-DiCDD	1:1	0.424	177.5
2,3-DiCDD	2:0	0.433	177.5
2,8-DiCDD	1:1		177.5
1,2,4-TriCDD	3:0	0.600	188.1
1,3,7-TriCdd	2:1		186.7
2,3,7-TriCDD	2:1	0.651	189.1
1,3,6,8-TCDD	2:2	0.813	196.2
1,3,7,9-TCDD	2:2	0.833	197.4
1,3,6,9-TCDD	2:2	0.852	
1,4,6,9-TCDD	2:2	0.896	202.2
1,2,4,7-TCDD	3:1	0.897	
1,2,4,8-TCDD			
1,3,7,8-TCDD	2:2	0.905	
1,2,4,6-TCDD	2:2	0.910	202.8
1,2,4,9-TCDD			
1,2,6,9-TCDD	2:2	0.972	203.6
1,2,3,6-TCDD	3:1	0.975	
1,2,3,4-TCDD	4:0	0.980	203.7
1,2,3,7-TCDD	3:1	0.985	203.9
1,2,3,8-TCDD			
2,3,7,8 TCDD	2:2	1.000	204.1
1,2,3,9-TCDD	3:1	1.01	
1,2,7,8-TCDD	2:2	1.03	
1,2,6,7-TCDD	2:2	1.04	205.6
1,2,8,9-TCDD	2:2	1.09	208.9
1,2,4,7,8 PeCDD	3:2	1.46	215.0
1,2,3,4,7 PeCDD	4:1	1.54	217.8
1,2,3,7,8 PeCDD	3:2	1.63	218.3
1,2,3,4,6 PeCDD	4:1		219.0
1,2,4,6,7,9 HxCDD	3:3	2.20	227.0
1,2,3,6,7,8 HxCDD	3:3	2.42	229.8
1,2,3,4,7,8 HxCDD	4:2	2.54	232.1
1,2,3,6,7,8 HxCDD	3:3	2.65	232.6
1,2,3,7,8,9 HxCDD	3:3	2.76	233.8
1,2,3,4,6,7,9 HpCDD	4:3	3.78	243.4
1,2,3,4,6,7,8 HpCCD	4:3	4.18	246.6
1,2,3,4,6,7,8,9 OCDD	4:4	6.76	250.0
1,2,6,8-TCDD	2:2	0.918	
1,4,7,8-TCDD	2:2	0.928	
1,2,7,9-TCDD	2:2	0.951	

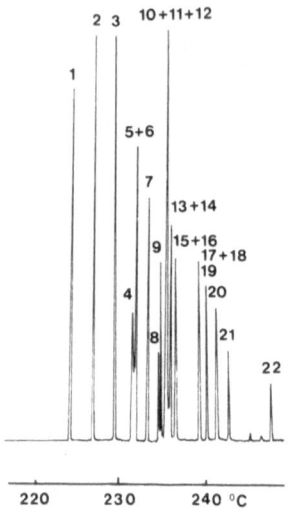

FIGURE 8.23. SIM chromatographic trace (*m/z* 320) showing elution of 22 tetrachlorodibenzo-*p*-dioxin isomers on 55-mm i.d. Silar-10 C WCOT column.

TABLE 8.24
Possible Combination of Three WCOT Columns Capable to Resolve 22 TCDD Isomers[a]

Isomer	Silar 10C	OV-17	OV-101
1,2,6,7	×	×	×
1,2,6,8	×		
1,2,6,9	×		
1,2,7,8			×
1,2,7,9			×
1,2,8,9	×	×	×
1,3,6,8	×	×	×
1,3,6,9	×	×	×
1,3,7,8	×	×	
1,3,7,9	×	×	×
1,4,6,9			
1,4,7,8	×		×
2,3,7,8	×		
1,2,3,6		×	
1,2,3,7			
1,2,3,8			
1,2,3,9	×		×
1,2,4,6			
1,2,4,7			
1,2,4,8			
1,2,4,9			
1,2,3,4			

[a] × indicates that isomers can be uniquely assigned on the WCOT column.

WCOT columns in PCDD analysis. However, methods for complete PCDD analysis have not been described that are applicable to very complex matrices, such as chemical wastes and landfill effluents. A summary of methods for determining mostly the 2,3,7,8-TCDD content of sample extracts is presented in Table 8.25. More methods entail use of either HRGC–MS or HRGC–HRMS–SIM.

PCDD have been the subject of much interest because of the high toxicity of the 2,3,7,8-TCDD isomer. This isomer was reported in fish from Vietnam collected in 1970 at levels as high as 800 ppt.[256] Fish samples collected between 1970 and 1977 from the Ohio River, Lake Michigan, the Hudson River (State of New York), the Connecticut River (State of Connecticut), the Tittabawassee River, Michigan, sediment samples from Lake Michigan (Illinois), a snapping turtle from the Hudson River (New York) and a seal from Baltic Sea were all analyzed for PCDD.[257]

TABLE 8.25
Summary of WCOT Applications in PCDD Analysis

WCOT column used[c]	References
18 m × 0.3-mm i.d., OV-61 WCOT column	250
22 m × 0.34-mm i.d., OV-17; TP:43500[a]	231
22 m × 0.32-mm i.d., OV-101; TP:42500[a]	231
22 m × 0.34-mm i.d., Silar 10c; TP:34000[a]	231
50 m × 0.36-mm i.d., OV-17; TP:140000[a]	217
50 m × 0.25 mm i.d., Silar 10C, TP:192000[a]	251
55 m × 0.37-mm i.d., OV-17; TP:152000[a]	251
55 m × 0.4-mm i.d., OV-101; TP:138000[a]	251
30 m × 0.25-mm i.d., SE-30; TP:113000	235
30 m × 0.25-mm i.d., OV-101; TP ~ 100000	252
50 m × 0.37-mm i.d., OV-17	253
40 m × 0.35-mm i.d., OV-101 at 220 °C isothermal	248
30 m × 0.25-mm i.d., OV-101; TP:108000[b]	247
30 m × 0.25-mm i.d., OV-1; TP:117000[b]	248
50 m × 0.25-mm i.d., CX-20M; TP:97000[a]	249
30 m × 0.5-mm i.d., (60 : 40) OV-17: Poly S-179: TP:30000	254
25 m × 0.2-mm i.d., SP-2100 at 215 °C isothermal	241
24 m × 0.2-mm i.d., OV-101 at 210 °C isothermal	
100 m × 0.25-mm i.d., Dexsil 300 at 220 °C isothermal	
30 m × Carbowax 20M, i.d. not given; at 240 °C isothermal	
28 m × i.d. not given, SP-2250 at 220 °C	
60 m × i.d. not given OV-101, temperature programmed from 190–	255
230 °C at 1 °C/min, hydrogen at 40 cm/sec	

[a] For octachlorodibenzo-p-dioxin.
[b] For C_{13}-hydrocarbon.
[c] TP is the number of theoretical plates per given length.

Isomers of tetra-, penta-, hexa-, hepta-, and octrachlorodibenzo-p-dioxins were found above the 500-ppt level in only two of these samples. However, fish samples from the Tittabawassee River at the Dow Dam in Midland, Michigan, contained approximately 1000 ppt of TCDD. In another study from the Love Canal, Buffalo area, 2,3,7,8-TCDD isomer concentration in sediment samples ranged from 0.9 up to 312 ng/g.[258]

8.9. CHLORINATED DIBENZOFURANS

The polychlorinated dibenzofurans (PCDF) are similar in structure to the polychlorinated dibenzo-1, 4-dioxins:

PCDD   PCDF

where x and y is $\geq$ 1 to 4. The formation of PCDF occurs primarily via the loss of ortho-H_2 and orto-HCl from chlorinated diphenyl ethers. These contaminants are found in polychlorinated phenols and in commercial PCB residues. Since these contaminants are formed as trace impurities in the preparation of pesticides and other industrial chemicals, analysis must permit the detection of congeners in ppt or ng/kg levels in environmental samples. In all, 135 PCDF isomers are possible, containing one to eight chlorine atoms.

The contamination of soil and water gives rise to concern about the appearance of these contaminants in the food supplies. In 1968 many people in Japan were intoxicated after they had used a cooking rice oil accidentally contaminated with a PCB-containing chlorinated dibenzofurans (Yusho oil). However, analysis of PCDF in the environment has been a formidable task for the environmental analytical chemists due to the ultratrace amounts at which they occur and the presence of interfering compounds in fatty tissues.[259] Highly effective cleanup procedures are required to remove PCDF congeners and enrich extracts containing PCDF.

A highly effective multiresidue cleanup and preconcentration procedure has been developed by Stalling.[260] The procedure may be summarized as follows:

1. Extract 100 g fish tissue with 1 l of methylene chloride after blending it first with four times its weight of Na_2SO_4.
2. After removal of methylene chloride by a rotary evaporator, the weight of the extracted oil is determined gravimetrically.
3. The oil is dissolved in a 1 : 1 (v/v) mixture of methylene chloride-cyclohexane. The total volume of the solvent mixture is adjusted to the concentration of 0.2 g of the oil per milliliter.
4. Aliquots (5 ml) are applied to a GPC column containing 70 g of SX-3 Bio-Beads Resin column (25 × 480 mm) using a 5-ml/min flow rate. The 155–300-ml eluate fraction is collected.
5. The sample fraction is reduced to near dryness by rotary evaporator, and all fractions are combined prior to its fractionation and transfer to gel permeation–cesium silicate–carbon-foam chromatography (CFC).
6. Chromatography (CFC) system for recovery of the PCDF. A total of 50 mg of AMOCO PX-21 carbon (d_p < 40 μm) was contained in the CFC column.
7. PCDF and planar molecule contaminants are eluted with 30 ml of toluene. The toluene is evaporated to a small volume (100 μl) for analysis by HRGC or GC/MS.

Polychlorinated dibenzofurans are the subject of growing concern in the ecosystem since they have been found in technical PCB mixtures,[261,262] chlorophenols,[250] and hexachlorobenzene.[264] All are chemical products that have been manufactured in large quantities and are recognized as environmental pollutants. Further, it is recognized that they can be generated from polychlorinated biphenyl ethers by photochemical reactions.

Problems involving quantitation of PCDF are largely related to the necessity of detecting very low concentrations, often in range of 1–10 ppt per congener (10^{-12} g/g). The development of a variety of sample cleanup techniques that permit PCDF isolation and preconcentration coupled with high-resolution GC–ECD or MS detectors can provide adequate detection and PCDF identification capability.[265,266,211] The completely specific identification of tetrachlorodibenzofuran congeners has not been reported to date. Largely this can be attributed to the lack of availability of TCDF isomers. The 38 positional isomers of TCDF have been synthesized by pyrolysis of specific polychlorinated biphenyls, ultraviolet photolysis of pentachlorodibenzofurans, and chlorinated of trichlorodibenzofurans by aromatic substitution.[267] The techniques involving HRGC have been reported.[231,253,257] Relative retention times for some chlorinated dibenzofurans and chlorinated dibenzo-p-dioxins are given in Table 8.26 and shown in Fig. 8.24.

The polychlorinated dibenzofurans have also been identified as residues in fish, sediment, Baltic seals, and Hudson River turtles[257] using negative ion CIMS in tandem with HRGC. Amounts of PCDF found in a fat portion of a snapping turtle and a grey Baltic seal were 6780 pg/ g (6.78 ppb) for the turtle and 77 pg/g (77 ppt) for the seal. The chromatogram shown in Fig. 8.25 indicates a distribution of homologues separated on the 50-m OV-17 glass WCOT column. The turtle fat was found contaminated by pentachloro- and hexachlorodibenzofurans and seal fatty tissue predominantely by hexachlorodibenzofurans. The occurrence and composition of residues of PCDF in sediments were also examined.[268]

Samples were column extracted using cyclohexane–methylene chloride (1 : 1). The extracts were then subjected to an enrichment step using sequential columns of potassium silicate, silica gel, cesium silicate, silica gel, and carbon dispersed on glass fibers. The residue profiles observed in sediment samples were similar in that the higher hepta- and octachloro-congener groups predominated, whereas in fish the tetrachloro- and pentachlorodibenzofurans were dominant.

TABLE 8.26
Relative Retention Times of Some CDF and PCDD Isomers

Isomer	RRT	$RT(t_R)[s]$
2,8-dichlorodibenzofuran	0.189	
3,6-dichlorodibenzofuran	0.307	
2,4,6-trichlorodibenzofuran	0.539	
2,3,7,8-tetrachlorodibenzofuran	0.799	
2,3,7,8-tetrachlorodibenzo-p-dioxin	1.000	396
1,2,4,7,8-pentachlorodibenzofuran	1.257	
1,2,4,6,7,9-hexachlorodibenzofuran	1.727	
1,2,3,4,7,8-hexachlorodibenzofuran	2.057	
1,2,3,4,6,7,9-heptachlorodibenzofuran	2.538	
Octachlorodibenzo-p-dioxin	3.542	
Octachlorodibenzofuran	3.644	1444

FIGURE 8.24. Chlorinated dibenzofurans and dibenzo-p-dioxins in Salmon oil (250 ppt). (1) 2,8-DCDF; (2) 3,6-DCDF; (3) 2,4,6-TrCDF; (4) 2,3,7,8-TCDD; (5) 2,3,7,8-TCDF; (6) 1,2,4,7,8-PCDF; (7) 1,2,4,6,7,9-HCDF; (8) 1,2,3,4,7,8-HCDD; (9) 1,2,3,4,6,7,9-HpCDF; (10) OCDD; (11) OCDF. OV-17 WCOT column 12 m × 0.2-mm i.d. Temperature programmed from 190 °C (hold 2 min) at 4 °C/min to 240 °C (hold 15 min). Carrier gas helium at 50 cm/sec.

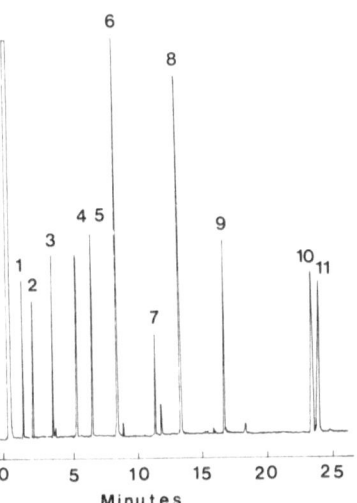

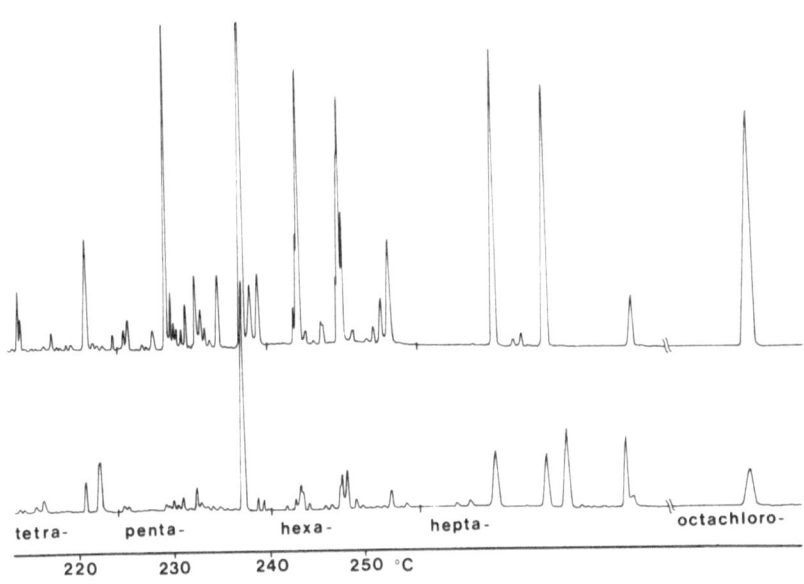

FIGURE 8.25. PCDF analysis of the snapping turtle fat extract on an OV-17 50-m WCOT column. Upper trace is the sample; lower trace is standard.

8.10. CHLORINATED PHENOLS AND RELATED COMPOUNDS

The increasing application of phenolic compounds, especially of chlorinated and nitrated phenols, which are used for manufacturing of various pesticides, disinfectants, and preservatives, requires an early and sensitive control of wastewaters and surface waters because of the low organoleptic threshold concentration levels in the environment.

8.10.1. Sample Preparation

It has been found that the stability of chlorinated phenols in water samples is low, and there is a need to preserve samples by acidification. Significant losses of chlorinated phenols cannot be prevented for more than 24 hr unless the residues are extracted into benzene and the extracts are stored in the dark at 0 °C.

Considering the various pathways through which chlorinated phenols can be degraded, sample handling, storage, and the period elapsed prior to their analysis are critical steps in sampling that should be well documented. It must be recognized that data obtained from samples subjected to prolonged storage before analysis could give a misleading assessment of the residue in the sample.

8.10.2. Extraction and Cleanup Procedures

Primary extraction procedures can be performed using various liquid–liquid solvent systems. An acid extraction medium is favored for all matrices except for air samples. Blending or homogenization is preferred to heating or distillation procedures. In general, hexane and benzene are preferred as the solvents, followed by diethyl ether, pentane, propanol, acetone, chloroform, dichloromethane, toluene, and acetonitrile. Extraction processes are governed by using the following criteria:

    (i) high solubility of the chlorinated phenols in extraction medium (high distribution coefficient, $K = C_E/C_w$) (Table 8.27) and
    (ii) small solubility of water in the extraction medium.

Table 8.27 indicates that diethyl ether–water extraction system for phenol should be preferred.

In Table 8.28 two extraction systems are shown for various phenols and chlorinated phenols and their distribution coefficients indicating their suitability for phenols. It is also recommended to use acetone–acid par-

TABLE 8.27
Distribution Coefficients of Phenol in Various Extraction Media

Extraction solvent	K value	Extraction solvent	K value
Diethyl ether	55	n-Butanol	18
2-heptanone	52	Methylcyclohexanol	18
3-heptanone	28	Benzene	2.1
Isoamylalcohol	28	Dichloromethane	1.3
Di-isopropyl ether	22	Cyclohexane	0.1

TABLE 8.28
Distribution Coefficients of Various Phenols in Two Different Systems

Compound	Diethylether–water K value	Diisopropyl ether–H$_2$O K value
Phenol	55	22
Resorcinol	5.4	1.5
o-Cresol	170	100
2,3-xylenol	260	190
o-Nitrophenol	170	110
o-Chlorophenol	130	90
m-Chlorophenol	320	
p-Chlorophenol	330	
2,3-dichlorophenol	870	
2,4-dichlorophenol	>000	
2,5-dichlorophenol	>1000	
2,6-dichlorophenol	580	
3,4-dichlorophenol	>1000	
3,5-dichlorophenol	>1000	
2,3,6-trichlorophenol	>1000	
2,4,5-trichlorophenol	>1000	
2,4,6-trichlorophenol	>1000	
2,3,4,5-tetrachloro-phenol	>1000	
2,3,4,6-tetrachloro-phenol	>1000	
Pentachlorophenol	>1000	
3-methyl-4-chlorophenol	>1000	
3-methyl-6-chlorophenol	590	
6-chlorothymol	>1000	

tition system, which is the most efficient solvent combination when a Soxhlet extraction system is employed.

Column chromatographic cleanup has been used in many cases employing Florisil, silica gel, ion exchange resins, and GPC; but in many cases where concentrated acid was used no further cleanup is required.

8.10.3. Derivatization of Phenols

Using fused-silica WCOT columns that are properly deactivated diminishes requirements for derivatization of phenols prior to their injection onto the column. However, when picogram quantities of chlorinated and nitrated phenols are injected, active sites can interact with the more polar phenols and they will tail or disappear completely. Derivatizing reagents employed for phenols include diazomethane, diazoethane, pentafluorobenzyl bromide, methyl iodide, acetic anhydride, 2,4-dinitro-1-fluorobenzene, and trimethylchlorosilane (TMS).

Symmetrical peaks and good yields were obtained for monochlorinated; 2,3-; 2,4-; 2,5-; 2,6-; 3,4-; and 3,5-dichlorophenols using TMS.[269] Acetylation can also be a suitable derivatization technique.[270] Relative retention times for some chlorinated, brominated, and nitrated phenol acetates are given in Table 8.29.

8.10.4. Qualitative and Quantitative Applications of WCOT Columns

Even when using WCOT columns, it is often difficult to resolve many pairs of isomeric phenols, but selective phases properly coated on the inert glass surface are capable of resolving all the phenols.[271] The gas chromatographic behavior of monohydric alkylphenols was studied on a tri(2,4-xylenyl) phoshate WCOT column modified with H_3PO_4. Data have been published on the retention characteristics of dihydric phenols[271] and chlorianted phenols[231] either as free phenols[272] on a 60 m × 0.5-mm i.d. stainless steel WCOT column coated with isodecylphthalate modified with H_3PO_4 and on two different fused-silica WCOT columns coated with SE-54 and FFAP using simultaneous injection onto two columns,[273] or as the methyl,[274] silyl,[269] or acetyl[275] derivatives. A 35 m × 0.35-mm i.d. WCOT operated isothermally at 130 °C can separate 15 of the 19 congeners.

The ECD detector has been utilized for chlorinated phenols. Analysis has also been based on GC/MS negative ion chemical ionization detection and electron impact ionization mass spectrometry.[277] The ECD responds to the different homologues of chlorinated phenols according to the chlorine content and a certain substitution pattern. This can be signifi-

TABLE 8.29[a]
Relative Retention Times for Halogenated Phenol Acetates Derivatives

| | Isothermal RRT | | Programmed from |
	150 °C	180 °C	95–180 °C at 3 °C/min
Chlrorphenol acetate			
2-monochloro	0.15	0.29	0.24
3-monochloro	0.15	0.30	0.27
4-monochloro	0.16	0.33	0.28
2,6-dichloro	0.20	0.33	0.34
2,5-dichloro	0.21	0.35	0.37
2,4-dichloro	0.21		0.38
3,4-dichloro	0.21	0.35	0.40
2,3-dichloro	0.23	0.36	0.43
3,5-dichloro	0.24	0.38	0.46
2,4,6-trichloro	0.27	0.41	0.49
2,3,6-trichloro	0.31		0.54
2,3,5-trichloro	0.33		0.56
2,4,5-trichloro	0.33	0.46	0.57
2,3,4-trichloro	0.38		0.64
3,4,5-trichloro	0.40		0.67
2,3,5,6-tetrachloro	0.50	0.61	0.73
2,3,4,6-tetrachloro	0.50	0.61	0.74
2,3,4,5-tetrachloro	0.64	0.71	0.83
Pentachloro	1.00	1.00	1.00
	(17.5 min)	(8.00 min)	(25.5 min)
Bromophenol acetates			
2-monobromo	0.18	0.32	0.32
3 monobromo	0.20	0.33	0.36
4-monobromo	0.20	0.33	0.37
2,6-dibromo	0.32	0.45	0.55
2,4-dibromo	0.35	0.48	0.61
Nitrophenol acetates			
2-mononitro	0.23	0.36	0.46
3-mononitro	0.29	0.41	0.58
4-mononitro	0.31	0.42	0.61

[a] WCOT 25 m × 0.35-mm i.d. coated with SE-30.

cantly eliminated while using derivatization with pentafluoropropionic or heptaflurobutyric anhydrides or by means of pentafluorobenzylbromide.[278] Techniques for confirmation of phenols rely heavily on both the resolving power and selectivity of different WCOT columns and the mass spectrometer as a selective detector.

Approximately 23,000 tons of pentachlorophenol are manufactured in the United States each year, and up to 8% of various hydroxyheptachloro- and hydroxyoctachlorodiphenyl ethers are formed as side-reaction products, besides 2,3,4,4,5,6-hexachloro-2,5-cyclohexadiene-1-

one. This impurity can produce a phenoxy radical at elevated temperature resulting in production of chlorinated dibenzo-p-dioxins, dibenzofurans, chlorinated diphenyl ethers, hydroxydiphenyl ethers, and chlorinated hydrobiphenyls.

Trichlorophenol has been identified in fish (fathead minnows) in addition to tetrachlorophenol, tetrachloroanisol, pentachlorophenol, and pentachloroanisol.[280] Pentachlorophenol has been found in fish exposed to the wastewater effluent from a wood-treating plant.[281] Pentachlorophenol has been found in fish from the Wabash River, Indiana.[282] In assessing the discharge of a wood-preservative plant releasing pentachlorophenol, the ratios of the 2,3,4,6-tetrachloro and 2,3,5,6-tetrachlorophenols had to be examined on a WCOT column 20 m × 0.25-mm i.d. coated with OV-101.[283] The pentachloroanisole has been found in fish from the Detroit River and the Arkansas River.[172] Lindström analyzed chlorinated phenols, guaiacols, and catechols from a pulp and paper factory after their derivatization with diazomethane and/or diazoethane.[284] The author used glass WCOT columns 25 m × 0.25-mm i.d. (liquid phase not given) and ECD or GC/MS–SIM detection. The musty taint associated with cured tobacco was analyzed, and the presence of chloroanisoles indicated as the cause of the taint using a 50 m × 0.75-mm i.d. WCOT steel column coated with Apiezon L.[285]

8.11. TOXAPHENE

Toxaphene represents an organochlorine insecticide (chlorinated camphene) employed in many countries all over the world for controling cotton pests. Toxaphene has been applied to crops mainly by aerial spraying. Toxaphene is used in lakes to control rough fish.

It is produced by the chlorination of camphene, which is carried out in the presence of uv light. The resulting product contains 67–69% chlorine. The similarity between congeners of technical chlordane and toxaphene makes the identification difficult when ECD is employed. The technical product has a wild terpene or pinelike odor. It undergoes dechlorination when heated above 120 °C or when exposed to uv light. Dechlorination may be accelerated in alkali media. Technical toxaphene is slightly soluble in water (3 ng/l), but highly soluble in turpentine, fuel oil, and kerosene. It is also soluble in acetone, benzene, toluene, and some chlorinated hydrocarbons such as tetrachloromethane and dichloroethylene.

Research interest into the chemical constituents of toxaphene has resulted from the toxicological interest to identify those isomers that exhibit toxic properties. There are many studies to resolve this problem, which is much more complex than PCB mixtures. At present more than 400 polychlorinated congeners are evident from fractionation by column chromatography.

It was shown that in addition to the presence of polychlorobornanes, polychlorinated dihydrocamphenes may also be present having the following structural formula:

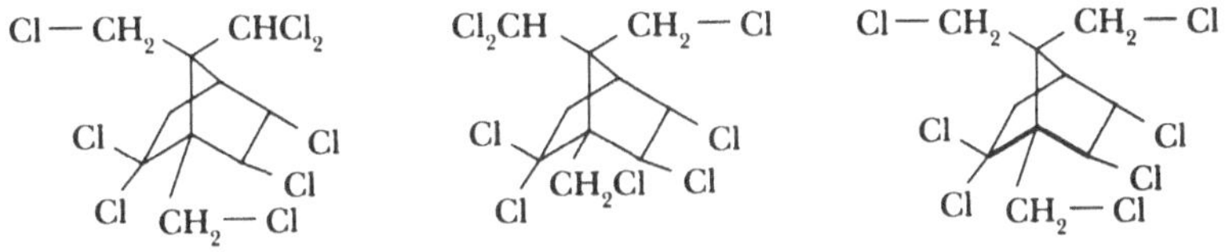

(I)   Toxicant A-1: 2,2,5-endo-6-exo-8,8,9,10-octachlorobornane
(II) Toxicant A-2: 2,2,5-endo-6-exo-8,9,9,10-octachlorobornane
(III) Toxicant B: 2,2,5-endo-6-exo-8,9,10-heptachlorobornane

The chemical complexity of toxaphene makes the metabolic study of the individual components very difficult.

8.11.1. Sample Preparation and Cleanup

Basically, characterization and quantitative analyses of toxaphene residues in fish, sediment, water, and wildlife samples follows most commonly employed multiresidue procedures for organochlorine compounds.[286] This consists of a solvent extraction step with petroleum ether–diethyl ether of a sample followed by GPC cleanup and fractionation on Florisil and silica gel columns. Gas chromatography with ECD or subsequent GC/MS–SIM can be used. Another alternate method requires extraction of ground tissue or sediment samples with dichloromethane followed by a cleanup of the extract by GPC and fractionation by Florisil and silica gel column chromatography. The extract is then characterized and quantified by WCOT column GC with ECD. Elimination of analytical interferences can be achieved by nitration with sulfuric–fumic HNO_3 acids (nitration mixture). Interferences from the PCB and DDT analogs can be removed effectively. However, nitration also reduces toxaphene concentration by about 5%. The flowchart of the method in a fish tissue multiresidue analysis is shown in Fig. 8.26.

8.11.2. WCOT Column Gas Chromatography for Toxaphene

Technical toxaphene represents a very complex mixture containing relatively small quantities of individual congeners. Various types of WCOT columns coated with OV-101[287,288] or SE-30 deactivated with Carbowax 20M and having phase ratio $\beta = 1000$ are used. The WCOT column (35 m $\times$ 0.30-mm i.d.) has SN = 48 for C_{12}–C_{13} with hydrogen as a carrier gas. The chromatogram obtained by the ECD on this column showed toxaphene + PCB fractions preseparated on a Florosil column.

An extensive study of toxaphene analysis by means of a WCOT column utilized 30 m $\times$ 0.22-mm i.d. fused-silica methylsilicone, 55 m

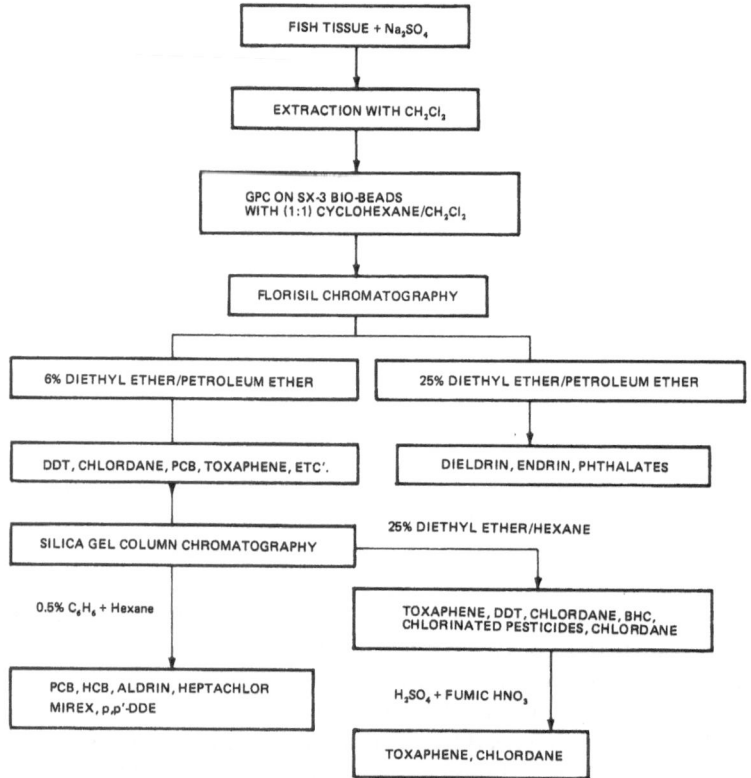

FIGURE 8.26. Schematic of the multiresidue analysis for toxaphene.

× 0.32-mm i.d. SE-54 fused-silica WCOT columns, and 20 m × 0.25-mm i.d. OV-17 borosilicate glass WCOT column.[286] Hydrogen was used as a carrier gas of 40 cm/sec and an initial column temperature of 140 °C was held for 10 min. Afterward, it was programmed to 250 °C at 2 °C/min and held at final temperature for 12.5 min. The injector temperature was 180 °C. The detector makeup gas was argon–methane (90 + 10) at 40 ml/min.

Onuska[289] used OV-1, OV-17, and OV-225 WCOT columns to produce the type of data shown in Fig. 8.27.

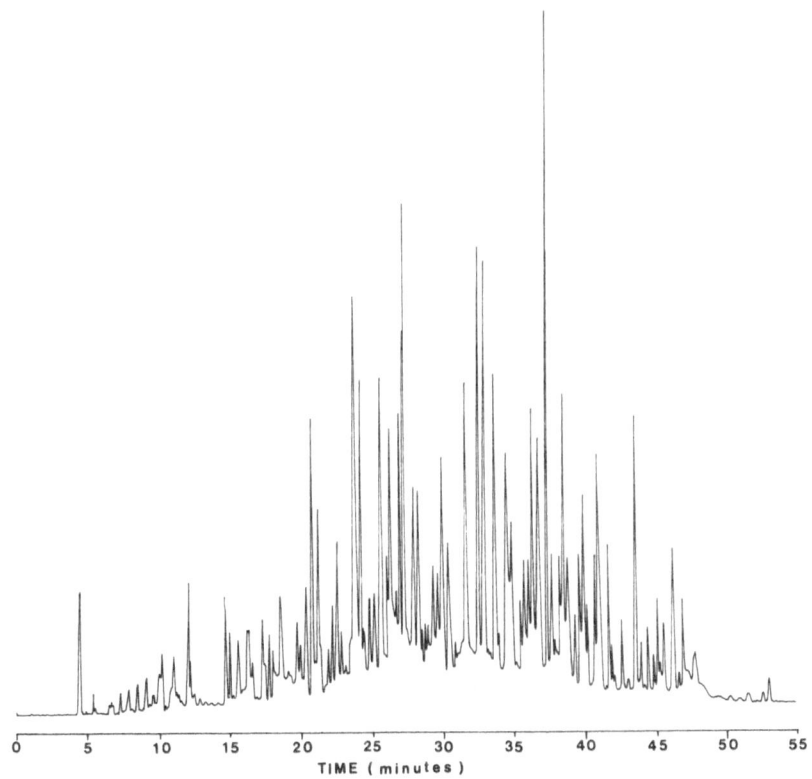

FIGURE 8.27. Toxaphene analysis on OV-1 FS-WCOT column, 25 m × 0.1-mm i.d. Cool on-column injector at 80 °C, hold for 1 min; programmed to 180 °C at 100 °C/min. WCOT column at 70 °C programmed to 240 °C at 4 °C/min final hold for 10 min.

254 CHAPTER EIGHT

8.11.3. Quantitative Analysis of Toxaphene

There is still no satisfactory methodology for quantifying toxaphene samples contained in environmental matrices. Usually, the WCOT chromatograms are quantified by running selected peak areas assignable to the unaltered toxaphene standard that did not coelute with peaks representing known interferences such as chlordanes, *trans*-nonachlor, *p,p'*-DDE, *p,p'*-DDD, and *cis*-nonachlor. The peaks representing toxaphene congeners can be identified by using the matching relative retention time index.[286] In this case *trans*-nonachlor is used for marking the elution range of 35.5 min, which is equal to an RRI value of 1.000 as the reference peak. Only 29 peaks, shown in Table 8.30, were considered for quantitation.

8.11.4. Toxaphene Analysis

Studies of air transport of chlorinated hydrocarbons indicate that nearly half of the total high molecular weight of chlorinated organic compounds in the air over the western North Atlantic were toxaphene congeners.[290] The presence of toxaphene in fish from Lake Vattern in Sweden is explained by airborne fallout of toxaphene.[291] Ballschmiter and Zell identified toxaphene in fish from a lake in the Tyrolian Alps and Ireland, but the sturgeon–savruga fish sample from the Caspian Sea had the highest value of toxaphene.[288] The most extensive survey of toxaphene in fish was run by the Columbia National Fisheries Laboratory.[292] They used GC/MS electron impact (EI) and chemical ionization techniques. Negative ionization mass spectrometry in contrast to EI mass spectra yields both greater sensitivity and less fragmentation. Residues in fish from Llano Grande Lake, Texas, Lake Michigan, and Baltic Sea seals exhibited toxaphene presence.

TABLE 8.30
Relative Retention Indices for Selected Toxaphene Peaks

Peak	RRI	Peak	RRI	Peak	RRI	Peak	RRI	Peak	RRI	Peak	RRI
1	1.036	6	1.140	11	1.194	16	1.275	21	1.345	26	1.461
2	1.043	7	1.154	12	1.207	17	1.293	22	1.365	27	1.476
3	1.089	8	1.158	13	1.237	18	1.297	23	1.387	28	1.483
4	1.099	9	1.170	14	1.253	19	1.322	24	1.390	29	1.510
5	1.125	10	1.176	15	1.263	20	1.331	25	1.412		

8.12. PHTHALATE ESTERS

The assessment of the true environmental level of phthalate esters pollution has been confused by difficulties in achieving blanks significantly lower than the environmental concentrations. Several authors have tackled this problem by simplifying the analytical methodology to eliminate the opportunity for laboratory cross contamination. Most of the published methods for the analysis and identification of phthalate esters involve WCOT column gas chromatography.

It has been shown[293] that phthalates, isophthalates, and terephthalate esters do not exhibit the same chromatographic behavior, but some of them, such as di(n-heptyl)- and (n-hexyl, n-octyl) phthalates, are not separated even on a WCOT column. However, they can be distinguished by GC/MS.

COOR

COOR

COOR — COOR
COOR

COOR

COOR

| Phthalate | Terephthalate | Isophthalate |

8.12.1. Qualitative Analysis

Retention indices of different phthalate esters on WCOT columns coated with SE-30 and OV-101 are presented in Table 8.31.

8.12.2. Sample Preparation and Cleanup

Water samples can be effectively extracted by means of 15% methylene chloride (DCM) in n-hexane; no differences were observed among the various pH conditions. It can be concluded that 15% dichloromethane-n-hexane should be used for extracting the wastewater, and recoveries are better than 94%. Samples can be stored at a temperature of 4 °C. Under these conditions no interferences are expected from residual chlorine up to the 2-ppm level. If Florisil cleanup is applied, the recoveries from Florisil or alumina are averaging 90% or better.

Water samples (1000 ml) are extracted with three times 60 ml DCM for each extraction. The combined extract is dried with Na_2SO_4 and evaporated to 5 ml in Kuderna–Danish evaporator and 90 ml of n-hexane

TABLE 8.31
Kovats Retention Indices of Phthalate Esters[a]

Compound	SE-30[b] 250 °C helium	OV-101, 230 °C nitrogen
Dimethyl phthalate		1453
Diethyl phthalate	1583	1581
Di(n-propyl) phthalate	1758	1756
Di(n-butyl) phthalate	1940	1937
Di(n-pentyl) phthalate	2122	2120
Di(n-hexyl) phthalate	2308	2305
Di(n-heptyl) phthalate	2497	2494
Di(n-octyl) phthalate	2685	
Di(n-nonyl) phthalate	2876	
Di(n-decyl) phthalate	3067	
Dimethyl terephthalate		1530
Dimethyl isophthalate		1512
Diethyl terephthalate	1650	1649
Diethyl isophthalate	1638	1638
Di(n-propyl) terephthalate	1851	1860
Di(n-propyl) isophthalate	1829	1828
Di(n-butyl) terephthalate	2060	2058
Di(n-butyl) isophthalate	2030	2025
Di(n-heptyl) terephthalate	2665	2661
Di(n-heptyl) isophthalate	2608	2605
Di(2-ethyl hexyl)phthalate	2509	
Dicylocohexyl phthalate	2475	

[a] From Ref. 294.

is added. The sample is preconcentrated to 2 ml. This extract is subjected to the Florisil cleanup procedure. The 100 ml of eluting solvent is collected containing all the phthalate esters. They are again preconcentrated to 1 ml volume for HRGC analyses.

Sediment samples were analyzed, and validation of the GC/MS–SIM analysis has been performed by Peterson and Freeman[295] (Table 8.32). They analyzed 30 g of wet sediment, and 160 ng of deuterated anthracene was used as an internal standard. The ultrasonic extraction was employed with methylene chloride for less than 5 min, and it was repeated three times with fresh solvent. Recoveries were higher than 90%. These authors used a 20 m × 0.27-mm i.d. WCOT column coated with SE-52. Splitless injections of 3 μl were analyzed by temperature programming from 150 to 275 °C at 7.5 °C/min. It was found possible to detect quantities as low as 100 pg of diethylhexyl phthalate in a sediment extract.

It is mandatory to identify potential sources of phthalate contami-

TABLE 8.32.
Characteristic Ions of Various Phthalate Esters

Compound	RRT[a]	m/z (abundance)
Dimethyl phthalate	0.46	163(100); 194(12)
Diethyl phthalate	0.68	149(100); 177(22); 222(3)
Diisobutyl phthalate	1.15	149(100); 223(6); 167(3)
Di-n-butyl phthalate	1.35	149(100); 233(4); 205(3)
Di-(2-ethylhexyl) phthalate	2.53	149(100); 167(38); 279(10)
Di-n-octyl phthalate	2.89	149(100); 279(4); 167(2)
Diisodecyl phthalate	3.00–3.4	149(100); 167(3-6); 307(2-4)
Di-n-decyl phthalate	3.61	149(100); 167(3); 307(3)

[a] RRT relative to the internal standard 10-d-anthracene.

nation in order to reduce their effect on the determination of phthalates. Blanks must be always run simultaneously with the sample.

8.13. NITROSAMINES

The N-nitrosamines (NA) are formed in the chemical reaction of secondary amines and nitrous anion. Secondary amines are present at low levels in environmental samples, and the nitrite anion is also a very common substance in the chemical industry. Many nitrosamines are potent animal and human carcinogens. Considerable research effort is underway to isolate, quantify, and identify these contaminants in foodstuffs. The analysis of uncooked meat, beverages, cosmetics, and tobacco condensates presents the greatest challenge for unexceptable levels of these toxic compounds.

8.13.1. Detection Systems

The most commonly used detection systems for nitrosamines by means of gas chromatography are the thermal energy analyzer (TEA),[296] the Coulson electrolytic conductivity detector,[297] and alkali flame ionization detector. None of these detectors has been utilized with WCOT column GC.

Only two detection systems: the photoionization detector (PID) and mass spectrometer in selected ion monitoring mode has been successfully applied to the analysis of nitrosamines. The use of the PID coupled to a WCOT column (50 m × 0.32-mm i.d. coated with Triton X305) shows sufficient separation power using a minimal cleanup as can be seen in Fig. 8.28.[298] The nitrosamines analyzed gave a linear response with a

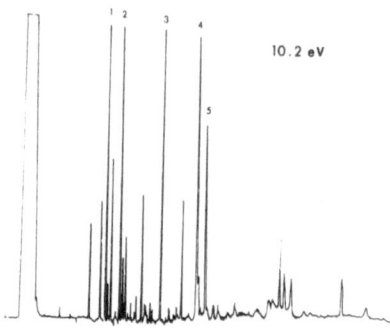

FIGURE 8.28. Analysis of nitrosamine standard mixture. (1) N,N-dimethylnitrosoamine; (2) N,N-diethylnitrosamine; (3) ethyl-butylnitrosoamine; (4) nitrosopiperidine; (5) nitrosopyrrolidine in CH₂Cl₂; approximately 6 ng/μl per component. Conditions are given in text.

lower detection limit of 50–100 pg with a 10.2-eV ionization source. It was pointed out that additional makeup gas is necessary for enhancing the sensitivity of the PID.

The mass spectrometer as a GC detector has been playing an important role in the identification and quantitative determination of N-nitroso compounds in trace analyses. In this connection it is useful to record a list of GC phases used in nitrosamine analysis.[299] The high-resolution MS–SIM on the nitrosamines has used HRGC for the separation of the components in the mixture. High-efficiency columns can reduce the possibility of interference between artifact peaks and peaks of interest.[300–302]

8.13.2. Applications

One of the first ions to be monitored by high-resolution SIM was the [NO]⁺ at m/z 30. However, this ion is neither specific nor a major ion. By monitoring the molecular ion or a characteristic ion at 10,000 mass resolution, it is possible to detect less than ppb of individual nitrosamines. The mass spectral fragmentation of 24 dialkyl-N-nitrosamines is given by Saxby.[303] An extensive survey[302] of various meat products using WCOT column GC–HRMS ($R = 4000$) indicated levels of 2–6 μg/kg for nitrosodimethylamine. The use of WCOT columns directly coupled to a high-resolution MS allowed the limit of detection for volatile N-nitrosamines to be lowered to $3 \cdot 10^{-13}$ g (300 fg).[304] Many applications for determining N-nitrosamines are provided in review articles by Gough.[305,306]

8.14. NONIONIC DETERGENTS

The use of synthetic surface-active agents each year is estimated at $4 \cdot 10^5$ metric tons. Basically, there are two types of nonionic surfactants such as polyoxyethylene mono-n-alkyl ethers, having the following formula:

$$RO(CH_2CH_2O)_n H$$

where R is an alkyl group and n is the number-average degree of polymerization and nonylphenol ethoxylates with one and two oxyethylene groups having the general formula:

$$H_{19}C_9C_6H_4O-(CH_2CH_2-O)_{n-1} CH_2CH_2-OH$$

where n can be 1,2,3,4. Technical 4-nonylphenol is a complex mixture of isomeric compounds with differently branched structural arrangements of the nonyl side chains. These isomers are present in technical surfactants of this type.

8.14.1. Extraction Procedures

The extraction of polyoxyethylene nonionic surfactants of the first type from water and sewage effluents at trace levels can be performed by 1,2-dichloroethane. This solvent appears to be suitable as a nonfractionating phase for extracting nonionic surfactants.[307] Also, methylene chloride and batch extraction in a separatory funnel and continous liquid–liquid extraction system for organic solvents heavier than water has been used.[308]

An extraction method used by the authors[309] employed a 100-ml sample extraction into chloroform, and 1-nonanol as an internal standard was added. The solvent was gently evaporated almost to dryness, and 50% hydrogen bromide in glacial acetic acid reagent was added. Ampoules were sealed and the reaction was carried out at 150 °C for 3 hr. The reaction products were extracted with carbon disulfide and analyzed by HRGC–ECD. The reaction products were identified as brominated alkanes from C_{10} to C_{15}. Ethylene or propylene glycol was converted quantitatively into 1,2-dibromoethylene and 1,2-dibromopropylene.

8.14.2. Qualitative and Quantitative Analysis

Quantitative determination of biodegradable nonionic surfactants of polyethoxylate-alkanol type showed that almost complete degradation of the alkanol part of the surfactant occurs, while the polyethoxylate portion degrades slowly and incompletely.[309] The yield of the extraction has been 95 ±5% and internal standard formed during reaction provides direct quantitative mean to the other brominated alkanes. The GC trace of the electron capture detector is a direct measure of the alkanoles.[310]

Quantitation of the number of moles of ethylene oxide can be calculated as:

$$n_E = \frac{\overline{M_A} \times A_E}{44(100 - A_E)}$$

where

$$A_E = \frac{A_{EB}}{(A_{EB} + \Sigma A_{RB})}$$

The peak areas (A) of the various alkyl bromides are employed to quantitate the proportions of the corresponding fatty alcohols, and hence the average molecular weight (M_A) of the alkanols.

Nonylphenolethoxylates can be quantified using an assumption that distribution of ethoxy groups is known.[308] NPEO$_1$ represents 14%; NPEO$_2$ is 42%; NPEO$_3$ is 30%; NPEO$_4$ is 12%; and NPEO$_5$ is 2% as established by FID. A gas chromatogram showing this separation is presented in Fig. 8.29.

8.15. MISCELLANEOUS SEPARATIONS

8.15.1. Separation of s-Triazines and Substituted Urea Herbicides

WCOT column gas chromatography has been applied to separate s-triazine herbicides and urea herbicides having the following formulas:

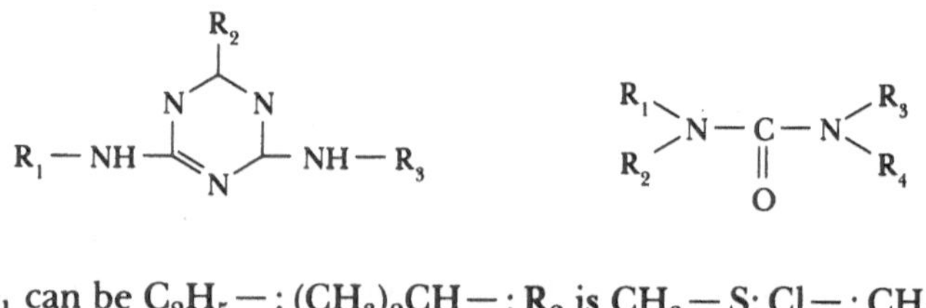

where R_1 can be C_2H_5-; $(CH_3)_2CH-$; R_2 is CH_3-S; $Cl-$; CH_3O; R_3 is $(CH_3)_2CH-$; $(CH_3)_2C(CN)-$; C_2H_5-; and R_4 is $H-$; CH_3-; C_4H_9-.

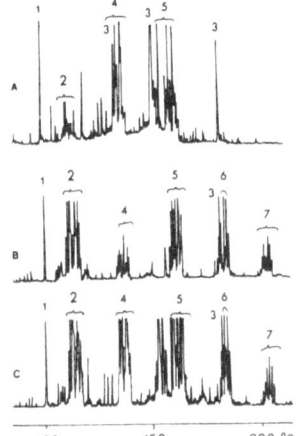

FIGURE 8.29. Nonylphenolethoxylate detergent analysis.
(A) Extract of secondary sewage effluent; (B) standard
mixture of nonylphenol and Marlophene 83; and (C) co-
injection of (A) and (B). (1) 2,4,6-triboromophenol; (2)
nonylphenols; (3) phthalates; (4) nonylphenolethoxylates
(NPEO) ($n = 1$); (5) NPEO ($n = 2$); (6) NPEO ($n = 3$);
(7) NPEO ($n = 4$); where n = number of ethoxy groups.

Two different WCOT columns were employed. For s-triazines a 22-
m $\times$ 0.25-mm-i.d. glass WCOT column coated with a 0.1-μm film of
SE-52 was used. Both ECD and NPSD detectors were used. The temperature-programming conditions from 60 to 110 °C at 30 °C/min, then
2 °C/min up to 220 °C. For urea herbicides, a 15 m $\times$ 0.3-mm i.d. glass
WCOT column coated with a 0.15-μm SE-52 film thickness was employed.
Nitrogen was used as a carrier gas. The column temperature was programmed from 70 to 90 °C at 30 °C/min, then at 8 °C/min to 210 °C.[311]

The quantitative analysis of s-triazine in model mixtures (such as
simazine, prometryn, atrazine, and propatine) were analyzed on Carbowax 20M as stationary phase WCOT column using FID and an alkali
flame detector. The concentration range of linear responses of both AFD
and FID were 1–200 ng and 10–2000 ng for the second detector.[314]

8.15.2. Volatile Chelates of Metal Ions

The very promising area of separating various metal ion chelates
such as trifluoroacetones, dialkyldithiocarbamates, di(trifluoroethyl) di-
thiocarbamates, and diisopropyldithiophosphates has been demonstrated
using WCOT columns.[312] Short WCOT-FS columns coated with SE-30,
Dexsil 300, SE-52, and SE-54 were employed. Variety of detectors such
as FID, ECD, flame photometric detector, and flameless nitrogen–phosphorus detector were used. Minimum detectable quantities for all chelates
are also reported.

Microwave plasma emission detection system with the cylindrical resonance cavity was applied as the element-selective detection system of volatile organometallic compounds containing V, Cr, Mn, Fe, Co, and Ni using fused-silica WCOT columns. The low picogram detection limits for most metals makes the microwave detection system very attractive for multicomponent multielement analyses.[313]

8.15.3. Odorous Compounds in Water, Fish, and Food

The three compounds most frequently identified as causing off flavors in water are (I) 2-methylisoborneol, (II) geosmin, and (III) β-cyclocitral:

(I)   (II)   (III)

These compounds are produced by actinomycetes and blue-green algae, and their threshold odor concentration is evident at the 10-ng/l level.[315,316] The analysis of water for taste and odor-causing compounds is usually undertaken when the problem is at its worst. Usually, dichloromethane is used in the extraction of the water (1.5 l). The extracts are concentrated to leave a small volume of 1–2 ml. This volume is quantitatively transferred to a graduated 4-ml threaded microvolumetric flask and further concentrated to 500 μl through a 25-cm Vigreux column. The concentrated sample is sealed in the distillation flask using a screw cap and silicone septum. Extracts can be analyzed with a GC system equipped with a WCOT column and FID or by GC/MS.

The chromatogram of Burlington, Ontario raw water sample shown in Fig. 8.30 confirms the presence of geosmin. Extracts were analyzed on a Carlo Erba 4160 GC in tandem with Finnigan MAT 311A mass spectrometer, equipped with a 60 m × 0.32-mm i.d. OV-1 cross-linked WCOT column using helium as carrier gas. The WCOT column was programmed from 140 °C at 7 °C/min to 280 °C and held 10 min. The cold on-column injection technique using 1-μl injection was used. The mass spectrum of geosmine is presented in Fig. 8.31.

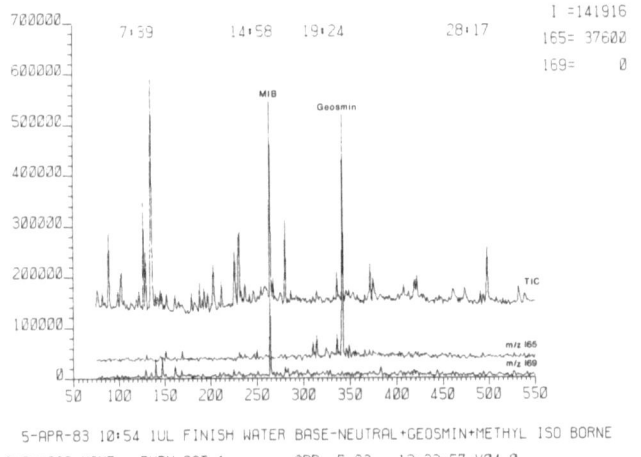

FIGURE 8.30. Burlington, Ontario water spiked with geosmine and 2-methyl-isoborneol (MIB).

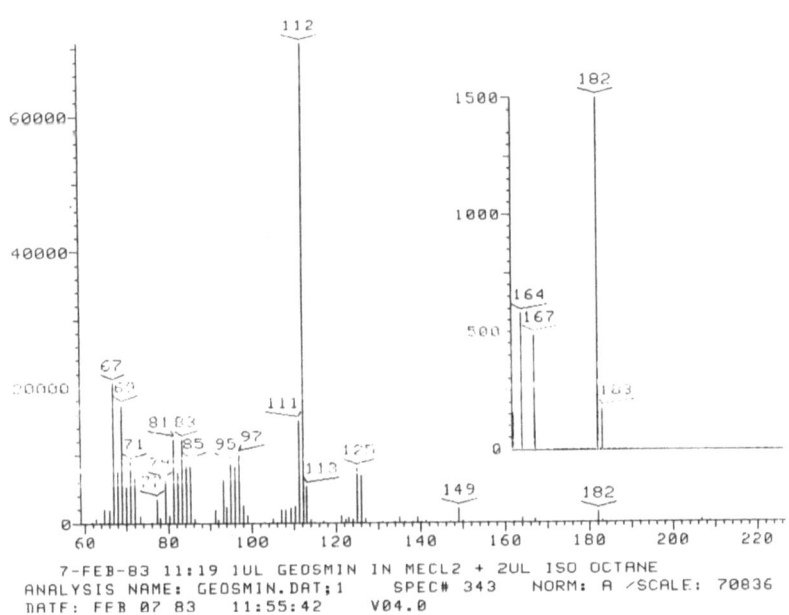

FIGURE 8.31. Electron impact mass spectrum of geosmine.

8.15.4. Fatty Acids

Environmental analytical chemists interested in applying WCOT columns to the research of lipid chemistry can benefit by observing some important factors.

The usual WCOT column injection devices can contribute to poor performance of WCOT columns. The volume of many injection devices is large in proportion to the WCOT column and sample size or carrier gas flow rate. The appropriate sample introduction onto the WCOT column can be sought in on-column injection. This term implies injection of a small sample containing fatty acids or their esters within the capillary tubing at the WCOT column temperature. Currently, cold on-column injection techniques are presenting the best choice.

The flame ionization detector, in addition to the mass spectrometer, is utilized as the principal detection system in lipid research. The procedure in the study of naturally occurring lipids involves saponification. The fatty acids are recovered and coverted to alkyl esters (usually methyl esters) by refluxing for 2 hr with 1% HCl and 20-fold molar excess of methanol or ethanol. The methyl (or ethyl) esters are separated from less than 1% of unesterified fatty acid by treatment with aqueous sodium carbonate or bicarbonate. Excessive concentrations of alkali and high temperatures in the saponification step should be avoided. Esterification may be carried out also with methanol containing 10% boron trifluoride. Heating for 5 min carries the reaction to completion.

The combination of diazomethane to give methyl esters and hexamethyl disilazane to give trimethylsilyl ethers or esters can provide volatile derivatives suitable for GC of a wide variety of substituted fatty acids.

The major types of fatty acids found in fish and animal fats are straight chain with some *iso* and *anteiso* and isoprenoid structures and some methyl-branched saturated fatty acids.[317] In the unsaturated fatty acids the fatty acids will be *cis* in structure, methylene interrupted, and have ω values of 3, 6, or 9. In animal lipids the types of fatty acids are limited in number.

Nonpolar stationary phases do not separate fatty acids well, especially related types such as $18:2\omega G$ and $18:3\omega 3$. In general, the introduction of a new double bond in a $\omega 6$-type acid to give a $\omega 3$-type acid does not significantly alter retention time on a nonpolar liquid phase. However, nonpolar phases have an advantage in that the even chain lengths of normal fatty acids do not overlap. This is illustrated with an analysis of Lake Ontario sediment samples from A.D. 1800 shown in Fig. 8.32. It is thus possible to determine the saturated fatty acids, the monounsaturated fatty acids, and certain polyunsaturated fatty acids as well as the com-

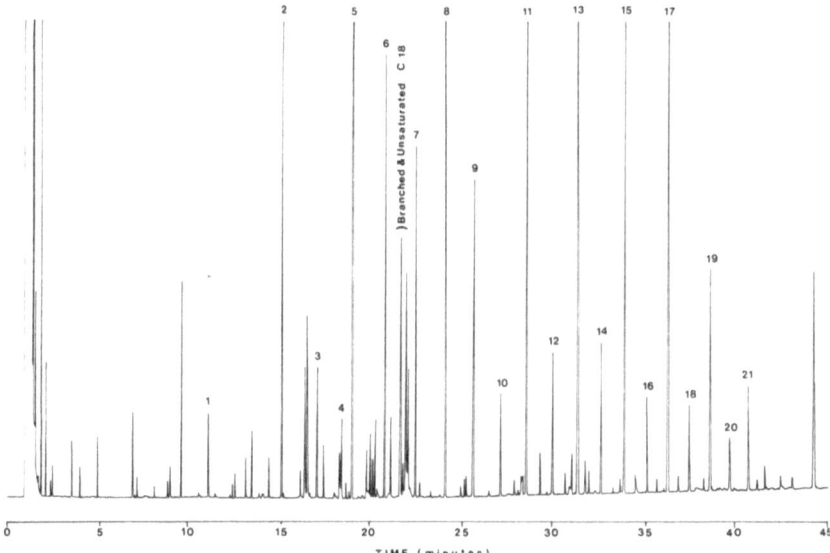

FIGURE 8.32. Chromatogram of fatty acid methyl esters extracted from Lake Ontario sediment dated A.D. 1800. (1) C12 : 0; (2) C14 : 0; (3) C15 : 0; (4) C16 : 1; (5) C16 : 0; (6) C17 : 0; (7) C18 : 0; (8) C19 : 0; (9) C20 : 0; (10) C21 : 0; (11) C22 : 0 (50 ng); (12) C23 : 0; (13) C24 : 0; (14) C25 : 0; (15) C26 : 0; (16) C27 : 0; (17) C28 : 0; (18) C29 : 0; (19) C30 : 0; (20) C31 : 0; (21) C32 : 0; on SE-30, 30 m × 0.32-mm i.d. FS-WCOT column programmed from 75 to 300 °C at 5 °C/min using cold on-column injection; FID at 325 °C. (Chromatogram courtesy of Dr. R. Bourbonier, Canada Centre for Inland Waters, Burlington, Ontario.)

position by totals of each chain length. The advent of more polar liquid phases represents the real breakthrough in the analysis of unsaturated fatty acids. Unfortunately, nearly all the polar phases that give efficient WCOT columns also give partial overlap of adjacent even chain lengths.

REFERENCES

1. U.S. Federal Reg. 44:69464-575.
2. "Sampling and Analysis Procedures for Survey of Industrial Effluents for Priority Pollutants," U.S. Environmental Protection Agency, Washington, D.C., March, 1977.
3. T. A. Bellar and J. J. Lichtenberg, *J. Am. Water Works Assoc.* **66**, 739–744 (1974).
4. W. L. Budde, J. W. Eichelberg, and P. Olynyk, "The Analysis of Trihalomethanes in Drinking Water Using Gas Chromatograph/Mass Spectrometry," April, 1980.
5. W. E. Coleman, R. D. Lingg, R. G. Melton, and F. C. Kopfler, in *Identification and Analysis of Organic Pollutants in Water* (L. H. Keith, ed.), Ann Arbor Science, Ann Arbor, Michigan, 1976, pp. 305–328.
6. B. S. Middleditch, S. R. Missler, and H. B. Hines, *Mass Spectrometry of Priority Pollutants*, Plenum, New York, 1981.

7. F. I. Onuska, *Can. Res.* **12**(2), 26–33 (1979).
8. R. R. Freeman, J. W. Pratt, Hewlett-Packard Appl. Note AN228-19, Avondale, PA.
9. A. R. Trussell, J. G. Moncur, F-Y. Lieu, and L. Y. C. Leong, *J. HRC & CC* **4**, 156–163 (1981).
10. F. I. Onuska and K. Terry, *Water Poll. Res. J. of Canada*, **17**, 103–115 (1983).
11. R. R. Freeman, T. A. Rooney, T. M. Przybylski, and L. H. Altmayer, Hewlett-Packard Appl. Note No. 83, 1979.
12. A. D. Sauter, L. D. Betowski, T. R. Smith, V. A. Strickler, R. G. Beimer, B. N. Colby, and J. E. Wilkinson, *J. HRC & CC* **4**(8), 366–383 (1981).
13. J. S. Warner, "Analytical Procedures for determining Organic Priority Pollutants in Municipal Sludge," U.S. Environmental Protection Agency, Washington, D.C., EPA-600/2, 80-030-EPA, March, 1980.
14. F. DeWalle and E. Chian, "Presence of Priority Pollutants in Sewage and Their Removal in Sewage Treatment Plants," US EPA Interim Report, May 1979, US-EPA, Cincinnati, OH.
15. J. W. Farrington, "Analytical Techniques for the Determination of Petroleum Contamination in Marine Organisms," U.S. Nat. Techn. Info. Service Report No. 766792/6, U.S. Government Printing Office, Washington, D.C., 1973.
16. M. Blumer, P. C. Blokker, E. B. Cowell, and E. J. Duckworth in *A Guide to Marine Pollution* (E. D. Goldberg, ed.), Gordon and Breach, New York, 1972.
17. A. Zsolnay, *Deep-Sea Res.* **20**, 923–925 (1973).
18. S. M. Ahmed, M. D. Beasley, A. C. Efromson, and R. A. Hites, *Anal. Chem.* **46**, 1856–1860 (1974).
19. E. R. Adlard, L. F. Creaser, and P. H. Matthews, *Anal. Chem.* **44**, 64–73 (1972).
20. D. V. Rasmussen, *Anal. Chem.* **48**, 1562–1566 (1976).
21. F. Berthou and M. P. Friocourt, *J. Chromatogr.* **219**, 393–402 (1981).
22. O. C. Zafiriou, *Estuar. Coast. Marine Sci.* **1**, 81–87 (1973).
23. M. P. Friocourt, Y. Gourmelun, F. Bertou, R. Cosson, and M. Marchand, Document CNEXO, No. 78/5752, Brest, France, pp. 617–631.
24. F. W. Karasek, O. Hutzinger, Res. and Dev. **26**, 40–46 (1975).
25. W. E. May, S. N. Chesler, S. P. Cram, B. H. Gump, H. S. Hertz, and D. P. Enagonio, *J. Chromatogr. Sci.* **13**, 535–540 (1975).
26. A. P. Bentz, *Anal. Chem.* **48**, 454A–472A (1976).
27. F. Zürcher and M. Thuer, *Environ. Sci. Technology* **12**, 838–843 (1978).
28. T. Kappeler and K. Wuhrmann, *Water Res.* **12**, 327–342 (1978).
29. B. Versino, H. Knoeppel, M. DeGroot, A. Peil, J. Poelman, H. Schauerberg, H. Vissers, and F. Geiss, *J. Chromatogr.* **122**, 373–388 (1976).
30. S. Thompson and G. Eglinton, *Mar. Pollut. Bull.* **9**, 133–136 (1978).
31. E. B. Overton, J. Bracken, and J. L. Laseter, *J. Chrom. Sci.* **15**, 169–173 (1977).
32. R. Jeltes, E. Burghard, T. R. Thijsse, and W. A. M. Tonkelaar, *Chromatographia* 430–437 (1977).
33. K. Grob, *Chromatographia* **8**, 423–433 (1975).
34. R. L. Miller and N. G. Johansen, Perkin-Elmer Paper No. 62, presented at Pittsburgh Conference, Atlantic City, 1982.
35. D. H. Desty and A. Goldup, *Chromatrography* (E. Heftmann, ed.), van Nostrand-Reinhold, New York, 1975, pp. 915–955.
36. R. Schoellner, K. Platzdasch, and G. Ulber, *Z. Chem.* **17**(9), 321–331 (1977).
37. P. I. Sanin, *Usp. Khim.* **45**(8), 1361–1394 (1976).
38. G. Kalmutchi, *Rev. Chem. (Bucarest)* **27**(1), 67–70 (1976).
39. *ASTM Manual on Hydrocarbon Analysis*, 3rd ed., American Society for Testing and Materials, Philadelphia, PA, 1978.

40. L. R. Hilpert, W. E. May, S. A. Wise, S. N. Chesler, and H. S. Hertz, *Anal. Chem.* **50**, 458–463 (1978).
41. S. Tomii and T. Futami, *Sekiyu Gakkai Shi* **19**, 499–506 (1976).
42. M. G. Bloch, R. B. Callen, and J. H. Stockinger, *J. Chromatogr. Sci.* **15**, 504–512 (1977).
43. H. Schultz, *Erdoel-Kohle* **30**, 182 (1977).
44. S. Rang, S. Kuningas, A. Orav, and O. Eisen, *Chromatographia* **10**, 55–64 (1977).
45. S. Rang, S. Kuningas, A. Orav, and O. Eisen, *J. Chromatogr.* **119**, 451–460 (1976).
46. S. Rang, S. Kuningas, A. Orav, and O. Eisen, *J. Chromatogr.* **128**, 53–58 (1976).
47. S. Rang, S. Kuningas, A. Orav, and O. Eisen, *J. Chromatogr.* **128**, 59–63 (1976).
48. S. Rang, S. Kuningas, A. Orav, and O. Eisen, *Chromatographia* **10**, 115–122 (1977).
49. F. M. Black, L. E. High, and J. E. Sigsby, *J. Chromatogr. Sci.* **14**, 257–260 (1976).
50. R. A. Cudney, E. G. Walther, and W. C. Malm, *J. Air. Poll. Control Assoc.* **27**, 468–470 (1977).
51. B. S. Middleditch and B. Basile, *Anal. Letters* **9**(11), 1031–1034 (1976).
52. W. O. Berry and P. J. Stein, *Bull. Environ. Contam. Toxicol.* **18**, 308–316 (1977).
53. W. Engelward and L. Wennrich, *Chromatographia* **9**, 540–547 (1976).
54. L. Sojak and J. A. Rijks, *J. Chromatogr.* **119**, 505–521 (1976).
55. L. Sojak, J. Janak, and J. A. Rijks, *J. Chromatogr.* **142**, 177–189 (1977).
56. K. Grob and G. Grob, *J. Chromatogr.* **90**, 303–313 (1974).
57. K. Grob, K. Grob, Jr., and G. Grob, *J. Chromatogr.* **106**, 299–317 (1975).
58. F. I. Onuska and K. Terry, Canada Centre for Inland Waters Internal Report, Burlington, Ontario, 1982.
59. W. W. Youngblood and M. Blumer, *Geochim. Cosmochim Acta* **39**, 1303–1314 (1975).
60. M. Blumer, *Chem. Geology* **16**, 245–256 (1975).
61. R. C. Pierce and M. Katz, *Environ. Sci. Technol.* **9**, 347–353 (1975).
62. *Manual of Analytical Methods*, National Inst. of Occup. Safety and Health, 2nd ed., Vol. 1, Cincinnati, OH, Dept. of Health, 1977.
63. R. C. Clark, Jr., and J. C. Findley, Proceedings of the Joint Conference on Prevention and Control of Oil Spills, American Petroleum Institute, Washington, D.C., pp. 161–172, 1973.
64. M. A. Acheson, R. M. Harrison, R. Perry, and R. A. Wellings, *Water Res.* **10**, 207–212 (1976).
65. R. M. Harrison, R. Perry, and R. A. Wellings, *Water Res.* **9**, 331–346 (1975).
66. R. E. LaFlame and R. A. Hites, *Geochim. et Cosmochim Acta* **42**, 289–303 (1978).
67. W. Giger and M. Blumer, *Anal. Chem.* **46**, 1663–1671 (1974).
68. G. Grimmer and H. Bohnke, *JAOAC* **58**, 725–733 (1975).
69. F. I. Onuska, A. W. Wolkoff, M. E. Comba, R. H. Larose, M. Novotny, and M. L. Lee, *Anal. Lett.* **9**, 451–460 (1976).
70. L. Kalas, A. Mudroch, and F. I. Onuska, in *Hydrocarbons and Halogenated Hydrocarbons in the Aquatic Environment* (B. K. Afghan and D. Mackay, eds.), Plenum, New York, 1979, pp. 567–576.
71. D. L. Vassilaros, P. W. Stocker, G. M. Booth, and M. L. Lee, *Anal. Chem.* **54**, 106–112 (1982).
72. I. Ignatiadis, J. M. Schmitter, and G. Guiochon, private communication.
73. A. Liberti, G. P. Cartoni, and V. Cantuti, *J. Chromatogr.* **15**, 141–148 (1964).
74. J. R. Wilmshurst, *J. Chromatogr.* **17**, 50–60 (1965).
75. V. Cantuti, G. P. Cantoni, A. Liberti, and A. G. Torri, *J. Chromatogr.* **17**, 60–65 (1965).
76. N. Caruno and S. Rossi, *J. Gas Chromatogr.* **5**, 103 (1967).
77. G. Alberini, V. Cantuti, and G. P. Cartoni, *Gas Chromatography 1966* (A. B. Littlewood, ed.), Institute of Petroleum, London, 1967, p. 258.
78. T. H. Gouw, I. M. Whitmore, and R. E. Jentoft, *Anal. Chem.* **42**, 1394–1399 (1970).

79. G. Grimmer and H. Boehnke, *Z. Anal. Chem.* **261**, 310 (1972).
80. M. L. Lee, M. Novotny, and K. D. Bartle, *Anal. Chem.* **48**, 1566–1572 (1976).
81. A. Bjorseth, J. Knutzen, and J. Skei, *Sci. Total Environ.* **13**, 71–86 (1979).
82. G. Schomburg, R. Dielmann, H. Borwitzky, and H. Husmann, *J. Chromatogr.* **167**, 337–354 (1978).
83. H. Borwitzky and G. Schomburg, *J. Chromatogr.* **170**, 99–124 (1979).
84. M. L. Lee, M. Novotny, and K. D. Bartle, *Anal. Chem.* **48**, 405–416 (1976).
85. M. L. Lee and R. A. Hites, *Anal. Chem.* **48**, 1890–1893 (1976).
86. K. Grob, Jr., and K. Grob, *Chromatographia* **10**, 250–255 (1977).
87. M. L. Lee, G. P. Prado, J. B. Howard, and R. A. Hites, *Biomed. Mass Spectrom.* **4**, 182–186 (1977).
88. R. H. Bieri, M. K. Cueman, C. L. Smith, and C. W. Su, *Int. J. Environ. Anal. Chem.* **5**, 293–310 (1978).
89. W. Giger and C. Schaffner, *Anal. Chem.* **50**, 243–249 (1978).
90. M. L. Lee, D. L. Vassilaros, W. S. Pipkin, and W. L. Sorenson, in *Trace Organic Analysis*, NBS Special Publ. No. 519. U.S. Government Printing Office, Washington, D.C., 1979, p. 731.
91. C. M. White, A. G. Sharkey, Jr., M. L. Lee, and D. L. Vassilaros, in *Polynuclear Aromatic Hydrocarbons* (P. W. Jones and P. Leber, eds.), Ann Arbor Science, Ann Arbor, Michigan, 1979, p. 261.
92. R. V. Schultz, J. W. Jorgensen, M. P. Maskarinec, M. Novotny, and L. J. Todd, *Fuel* **58**, 783 (1979).
93. S. G. Wakeham, C. Schaffner, and W. Giger, *Geochim. Cosmochim. Acta* **44**, 403–414 (1980).
94. M. L. Lee, C. Willey, R. N. Castle, and C. M. White, Proceedings of the 4th International Symposium on PAHs, Columbus, Ohio, October, 1979.
95. P. E. Strup, J. E. Wilkinson, and P. W. Jones, in *Carcinogenesis*, Vol. 3 (P. W. Jones and R. J. Freudethal, eds.), Raven, New York, 1978, p. 131.
96. M. E. Snook, R. F. Severson, H. C. Higman, R. F. Arrendale, and O. T. Chortyk, in *Polynuclear Aromatic Hydrocarbons* (P. W. Jones and P. Leber, eds.), Ann Arbor Science, Ann Arbor, Michigan, 1979, p. 231.
97. A. Bjorseth and G. Lunde, *Atmos. Environ.* **13**, 45 (1979).
98. A. Bjorseth and G. Eklund, *J. HRC & CC* **2**, 22–26 (1979).
99. A. Bjorseth, *Anal. Chim. Acta* **94**, 21–27 (1977).
100. G. Lunde and A. Bjorseth, *Nature* **268**, 518–519 (1977).
101. R. C. Lao and R. S. Thomas, in *Polynuclear Aromatic Hydrocarbons* (P. W. Jones and P. Leber, eds.), Ann Arbor Science, Ann Arbor, Michigan, 1979, p. 429.
102. T. Doran and N. G. McTaggart, *J. Chromatogr. Sci.* **12**, 715 (1974).
103. F. S. C. Lee, T. J. Prater, and F. Ferris, in *Polynuclear Aromatic Hydrocarbons* (P. W. Jones and P. Leber, eds.), Ann Arbor Science, Ann Arbor, Michigan, 1979, p. 83.
104. R. L. Hanson, R. E. Royer, R. L. Carpenter, and G. J. Newton, in *Polynuclear Aromatic Hydrocarbons* (P. W. Jones and P. Leber, eds.), Ann Arbor Science, Ann Arbor, Michigan, 1979, p. 3.
105. W. H. Griest, B. A. Tomkins, J. L. Epler, and T. K. Rao, in *Polynuclear Aromatic Hydrocarbons* (P. W. Jones and P. Leber, eds.), Ann Arbor Science, Ann Arbor, Michigan, 1979, p. 395.
106. L. Sucre, W. Jennings, G. L. Fisher, O. G. Raabe, and J. Olechno, in *Trace Organic Analysis*. NBS Spec. Publ. 519, U.S. Government Printing Office, Washington, D.C., 1978, p. 109.
107. L. Blomberg and J. Wannman, *J. Chromatogr.* **148**, 379–387 (1978).

108. R. F. Severson, W. S. Schlotzhauer, O. T. Chortyk, R. F. Arrendale, and M. E. Snook, in *Polynuclear Aromatic Hydrocarbons* (P. W. Jones and P. Leber, eds.), Ann Arbor Science, Ann Arbor, Michigan, 1979, p. 277.

109. S. Saluste, J. Klesment, and S. Kivirahk, *Pesti NSV Tead. Akad. Toim. Keem.* **28**, 7 (1979).

110. F. I. Onuska and M. E. Comba, *J. Chromatogr.* **126**, 133–145 (1976).

111. D. Brocco, V. Cantuti, and G. P. Cantoni, *J. Chromatogr.* **49**, 66–69 (1970).

112. D. Brocco, A. Cimmino, and M. Possanzini, *J. Chromatogr.* **84**, 371–377 (1973).

113. L. T. Mariich and F. K. Lenkevicz, *Koks. Khim.* **5**, 34–42 (1972).

114. E. R. Adlard, L. F. Creasen, and P. H. D. Matthews, *Anal. Chem.* **44**, 207–212 (1972).

115. L. Blomberg, J. Bujten, J. Gawozik, and T. Wannman, *Chromatographia* **11**, 521 (1978).

116. G. Grimmer, J. Jacob, and K. W. Naujack, *Z. Anal. Chem.* **306**, 345–355 (1981).

117. E. Balfanz, J. Koenig, W. Funcke, and J. Romanowski, *Fresenius Z. Anal. Chem.* **306**, 340–346 (1981).

118. G. Gimmer, K. W. Naujack, and D. Schneider, *Fresenius Z. Anal. Chem.* **311**, 475–484 (1982).

119. M. L. Lee, D. L. Vassilaros, C. M. White, and M. Novotny, *Anal. Chem.* **51**, 768–777 (1979).

120. M. L. Lee and B. W. Wright, *J. Chromatogr. Sci.* **18**, 345–358 (1980).

121. W. Giger, C. S. Schaffner, *Adv. in Org. Geochem.* p. 375 (1975).

122. W. Giger, M. Reinhard, and C. S. Schaffner, *Vom Wasser* **43**, 343–358 (1974).

123. P. W. Jones, *Proceedings of the Analytical Division of the Chemical Society*, **15**, 158 (1978).

124. D. J. Hallett, R. J. Norstrom, F. I. Onuska, and M. E. Comba, *Proceedings of the 2nd International Symposium on Glass Capillary Chromatography*, Hindelang/Allgau, Germany, pp. 115–125 (1977).

125. J. M. Schmitter, I. Ignatiadis, and G. Guiochon, *J. Chromatogr.* **248**, 203–216 (1982).

126. G. Alexander and I. Hazai, *J. Chromatogr.* **217**, 19–38 (1981).

127. J. M. Schmitter, H. Colin, J. L. Excoffier, P. J. Arpino, and G. Guiochon, *Anal. Chem.* **54**, 769–772 (1982).

128. M. Dong, I. Schmeltz, E. Jacobs, and D. Hoffman, *J. Anal. Toxicol.* **2**, 21 (1978).

129. W. Cauthreels and K. van Canwenberghe, *Atm. Environ.* **10**, 447–457 (1976).

130. M. Novotny, R. Kump, F. Merli, and L. J. Todd, *Anal. Chem.* **52**, 401–406 (1980).

131. J. F. McKay, J. M. Weber, and D. R. Latham, *Anal. Chem.* **48**, 891–898 (1976).

132. B. A. Tomkins and C.-H. Ho, *Anal. Chem.* **54**, 91–96 (1982).

133. P. Burchill, A. A. Herod, and E. Prichard, *J. Chromatogr.* **246**, 271–295 (1982).

134. C. Patriankos and D. Hoffman, *J. Anal. Toxicol.* **3**, 150–154 (1979).

135. D. W. Later, M. L. Lee, and B. W. Wilson, *Anal. Chem.* **54**, 117–123 (1982).

136. J. Jaeger, *J. Chromatogr.* **152**, 575–578 (1978).

137. C. Y. Wang, M.S. Lee, C. M. King, and P. O. Warner, *Chemosphere* **9**, 83–89 (1980).

138. D. Schuetzle, F. S.-C. Lee, T. L. Parater, and S. B. Tejada, *Int. J. Environ. Anal. Chem.* **9**, 93–144 (1981).

139. W. L. Fitch, E. T. Everhart, and D. H. Smith, *Anal. Chem.* **5**, 2122–2126 (1978).

140. Th. Ramdahl and K. Urdal, *Anal. Chem.* **54**, 2256–2260 (1982).

141. H. J. Coleman, J. E. Dooley, D. E. Hirsch, and C. J. Thompson, *Anal. Chem.* **45**, 1724–1737 (1973).

142. M. Ogata and Y. Miyake, *J. Chromatogr. Sci.* **18**, 594–605 (1980).

143. M. Ogata and Y. Miyake, *Water Res.* **15**, 257–266 (1981).

144. M. L. Lee, D. L. Vassilaros, and D. W. Later, *Int. J. Environ. Anal. Chem.* **11**, 251–262 (1982).

145. M. Ogata, Y. Miyake, K. Fuhisawa, and Y. Yoshida, *Bull. Environ. Contam. Toxicol.* **25**, 130–135 (1980).

146. F. Berthou, Y. Gourmelun, Y. Dreano, and M. P. Friocourt, *J. Chromatogr.* **203**, 279–292 (1981).
147. D. W. Kuehl, *Chemosphere* **10**, 231–242 (1981).
148. H. Steinwandter, *Fresenius Z. Anal. Chem.* **304**, 137–140 (1980).
149. S. Goeb, H. Parlar, and F. Korte, *J. Agr. Food Chem.* **25**, 1224–1227 (1977).
150. M. P. Yurawecz, P. A. Dreifuss, and L. R. Kamps, *J. Assoc. Off. Anal. Chem.* **59**, 552–558 (1976).
151. G. Yip, *J. Assoc. Off. Anal. Chem.* **59**, 559–561 (1976).
152. M. P. Yurawecz, *J. Assoc. Off. Anal. Chem.* **62**, 36–40 (1979).
153. M. P. Yurawecz and J. A. G. Roach, *J. Assoc. Off. Anal. Chem.* **61**, 26–31 (1978).
154. D. W. Kuehl, E. N. Leonard, K. J. Welch, and G. D. Veith, *J. Assoc. Off. Anal. Chem.* **63**, 1238–1244 (1980).
155. H. R. DeLeon, M. A. Maberry, E. B. Overton, C. K. Rashke, P. C. Remele, S. F. Steele, V. L. Warren, and J. L. Leseter, *J. Chromatogr. Sci.* **18**, 85–88 (1980).
156. J. Barkely, J. Bunch, J. T. Bursey, N. Castillo, S. D. Cooper, J. M. Davis, M. D. Erickson, B. S. H. Harris, M. Kirkpatrick, L. C. Michael, S. P. Parks, E. D. Pellizzari, M. Ray, D. Smith, K. B. Tomer, R. Wagner, and R. A. Zweidinger, *Biomed. Mass Spectrom.* **7**, 139–147 (1980).
157. S. Dmochenwitz and K. Ballschmiter, *Fresenius Z. Anal. Chem.* **310**, 6–12 (1982).
158. M. Oehme and H. Stray, *Fresenius Z. Anal. Chem.* **311**, 665–673 (1982).
159. K. L. E. Kaiser, *Science* **185**, 523–525 (1974).
160. R. J. Norstrom, H. T. Won, M. V. H. Holdrinet, P. G. Calway, and C. D. Naftel, *J. Assoc. Off. Anal. Chem.* **63**, 37–42 (1980).
161. R. J. Norstrom, D. J. Hallett, F. I. Onuska, and M. E. Comba, *Environ. Sci. Technol.* **14**, 860–866 (1980).
162. F. I. Onuska, M. E. Comba, and J. L. Coburn, *Anal. Chem.* **52**, 2272–2275 (1980).
163. D. W. Kuehl, *Anal. Chem.* **49**, 521–523 (1977).
164. J. L. Laseter, J. R. DeLeon, and P. C. Remele, *Anal. Chem.* **50**, 1169–1172 (1978).
165. R. S. Harless, D. E. Harris, G. W. Sovocool, R. D. Zehr, N. K. Wilson, and E. O. Oswald, *Biomed. Mass Spectrom.* **5**, 232–237 (1978).
166. F. I. Onuska and M. E. Comba, *J. HRC & CC* **1**, 209–210 (1978).
167. F. Y. Saleh and G. F. Lee, *Environ. Sci. Technol.* **12**, 297–301 (1978).
168. B. G. Oliver and K. D. Bothen, *Anal. Chem.* **52**, 2066–2069 (1980).
169. D. J. Hallett, R. J. Norstrom, F. I. Onuska, and M. E. Comba, *Chemosphere* **11**, 277–285 (1982).
170. M. Morita and G. Ohi, *Chemosphere* **10**, 839–842 (1978).
171. G. A. Eiceman, R. E. Clement, and F. W. Karasek, *Anal. Chem.* **51**, 2343–2350 (1979).
172. G. D. Veith, D. W. Kuehl, E. N. Leonard, F. A. Puglisi, and A. E. Lemke, *Pest. Monitor. J.* **13**, 1–11 (1979).
173. W. F. Ten Berge and M. Hillebrand, *J. Sea Res.* **8**, 361–368 (1974).
174. E. B. Ofstad, G. Lunde, K. Martinsen, and B. Rugg, *Sci. Total Environ.* **10**, 219–230 (1978).
175. F. I. Onuska, A. Mudroch, and K. Terry, *J. Great Lakes Res.* **9**(2), 169–182 (1983).
176. J. Jan, *Chemosphere* **9**, 165–167 (1980).
177. H. Buchert, S. Bihler, P. Schott, H. P. Roeper, H. J. Pachur, and K. Ballschmiter, *Chemosphere* **10**, 945–956 (1981).
178. F. W. Crow, A. Bjorseth, K. T. Knapp, and R. Bennett, *Anal. Chem.* **53**, 619–625 (1981).
179. F. I. Onuska, CCIW Internal Report, "Chlorinated Benzenes," 1982.
180. K. Ballschmiter, Ch. Scholz, H. Burchert, M. Zell, K. Figge, K. Polzhofer, and H. Hoerschelmann, *Fresenius Z. Anal. Chem.* **309**, 1–7 (1981).

181. D. W. Kuehl, H. L. Kopperman, G. D. Veith, and G. E. Glass, *Bull. Environ. Contam. Toxicol.* **16**, 127–132 (1976).
182. G. Lunde and A. Bjorseth, *Sci. Total Environ.* **8**, 241–246 (1977).
183. L. Kahn and C. H. Wayman, *Anal. Chem.* **36**, 1340–1343 (1964).
184. N. V. Brodtman, Jr., and W. E. Koffskey, *J. Chromatogr. Sci.* **17**, 97–110 (1979).
185. M. Suzuki, Y. Yamato, and T. Watenabe, *Environ. Sci. Technol.* **11**, 1109–1113 (1977).
186. P. E. Mattson and S. Nygren, *J. Chromatogr.* **124**, 265–275 (1976).
187. J. J. Franken, in *Proceedings of the 7th International Symposium on Chromatography and Electrophoresis*, Academic, Brussels, 1972, p. 441.
188. H. Tausch and N. Zash, SGAE-Berichte No. 2635, BL-177/76, Seiberdorf, Austria, 1976.
189. A. Soedergren, *Oikos* **23**, 30–41 (1972).
190. D. L. Stalling, R. C. Tindle, and J. L. Johnson, *J. Assoc. Off. Anal. Chem.* **55**, 32–45 (1972).
191. A. Soedergren, *Bull. Environ. Contam. Technol.* **10**, 116–119 (1973).
192. J. Krupčik, J. Križ, D. Prušová, P. Suchanek, and Z. Červenka, *J. Chromatogr.* **142**, 797–807 (1977).
193. F. Onuska and M. E. Comba, in *Hydrocarbons and Halogenated Hydrocarbons in the Aquatic Environment* (B. K. Afghan and D. Mackay, eds.), Plenum, New York, 1980, pp. 285–302.
194. F. I. Onuska, R. J. Kominar, and K. Terry, *Proceedings of the 5th International Symposium on Capillary Chromatography, 1983*, Riva del Garda, Italy, 1983.
195. H. J. Neu, M. Zell, and K. Ballschmiter, *Fresenius Z. Anal. Chem.* **293**, 193–200 (1978).
196. K. Ballschmiter and M. Zell, *Fresenius Z. Anal. Chem.* **302**, 20–31 (1980).
197. S. Jensen and G. Sundstrom, *Ambio* **3**, 70–76 (1974).
198. M. A. Moseley and E. D. Pellizzari, *J. HRC & CC* **5**, 404–412 (1982).
199. F. Riedo, D. Fritz, G. Tarjan, and E. Kovats, *J. Chromatogr.* **126**, 63–83 (1976).
200. M. Zell, H. J. Neu, and K. Ballschmiter, *Chemosphere* **2/3**, 69–76 (1977).
201. M. D. Mullin and J. C. Filkins, LABCON 1981, Chicago, IL, 1981.
202. V. Zitko, O. Hutzinger, and S. Safe, *Bull. Environ. Contam. Toxicol.* **6**, 160–163 (1971).
203. W. H. Dennis and W. J. Cooper, Techn. Report 7702, U.S. Army Medical Corps Bioengineering Research Laboratory, Fort Detrick, 1977.
204. O. Berg, P. Diosady, and G. Rees, *Bull. Environ. Contam. Toxicol.* **7**, 338–347 (1972).
205. H. L. Crist and R. F. Moseman, *J. Assoc. Off. Anal. Chem.* **60**, 1277–1281 (1977).
206. T. R. Schwartz, D. L. Stalling, J. D. Petty, J. W. Hogan, B. K. Marlow, R. D. Campbell, and R. L. Little, American Chemical Society Symposium, Kansas City, Missouri, 1982.
207. W. D. Dunn, E. Johannson, D. L. Stalling, T. R. Schwartz, B. K. Marlow, J. D. Petty, and J. W. Hogan, American Chemical Society Symposium, Kansas City, Missouri, 1982.
208. T. Cairns and E. G. Siegmund, *Anal. Chem.* **53**, 1599–1603 (1981).
209. G. P. Martelli, M. G. Castelli, and R. Fanelli, *Biomed. Mass. Spectrom.* **8**, 347–350 (1981).
210. L. J. Carter, *Science* **192**, 240 (1976).
211. T. J. Farrell, *J. Chromatogr. Sci.* **18**, 10–17 (1980).
212. F. I. Onuska, unpublished results.
213. L. H. Wright, R. G. Lewis, H. L. Crist, G. W. Sovocol, and J. M. Simpson, *J. Anal. Toxicol.* **2**, 76–79 (1978).
214. F. W. Karasek and F. I. Onuska, *Anal. Chem.* **54**, 309A–315A (1982).
215. P. W. Albro and B. J. Corbett, *Chemosphere* **6**, 381–386 (1977).
216. H. R. Buser, *Anal. Chem.* **49**, 918–922 (1977).
217. H. R. Buser and C. Rappe, *Chemosphere* **7** 199–211 (1978).

218. R. E. Clement, G. A. Eiceman, F. W. Karasek, W. D. Bowers, and M. L. Parsons, *J. Chromatogr.* **189**, 53–59 (1980).

219. W. B. Crummett and R. H. Stehl, *Environ. Health Persp.* **5**, 15–25 (1973).

220. S. S. Cutié, *Anal. Chim. Acta* **123**, 25–31 (1981).

221. R. A. Hummel, *J. Agric. Food Chem.* **25**, 1049–1053 (1977).

222. L. L. Lamparski, N. H. Mahle, and L. A. Shadoff, *J. Agric. Food Chem.* **26**, 1113–1116 (1978).

223. D. W. Kuehl and R. C. Dougherty, *Adv. Mass Spectrom.* **8** (1980).

224. M. L. Langhorst and L. A. Shadoff, *Anal. Chem.* **52**, 2037–2044 (1980).

225. L. L. Lamparski, T. J. Nestrick, and R. H. Stehl, *Anal. Chem.* **51**, 1453–1458 (1979).

226. H. R. Buser and H. P. Bosshardt, *J. Chromatogr.* **90**, 71–77 (1974).

227. H. R. Buser and H. P. Bosshardt, *J. Assoc. Off. Anal. Chem.* **59**, 562–569 (1976).

228. C. D. Pfeiffer, T. J. Nestrick, and C. W. Kocher, *Anal. Chem.* **50**, 800–804 (1978).

229. T. Ramstad, N. H. Mahle, and R. Matalon, *Anal. Chem.* **49**, 386–390 (1977).

230. E. C. Villanueva, R. W. Jennings, V. W. Burse, and R. D. Kimbrough, *J. Agric. Food Chem.* **23**, 1089–1091 (1975).

231. H. R. Buser, *Anal. Chem.* **48**, 1553–1557 (1976).

232. A. DiDomenico, F. Merli, L. Boniforti, I. Camoni, A. DiMuccio, F. Jaggi, L. Vergori, G. Colli, G. Elli, A. Gorni, P. Grassi, G. Inveruizzi, A. Jemma, L. Luciani, F. Cattabeni, L. DeAngelis, G. Galli, C. Chiabrando, and R. Fanelli, *Anal. Chem.* **51**, 735–740 (1979).

233. J. J. Ryan and J. C. Pilon, *J. Chromatogr.* **197**, 170–180 (1980).

234. P. W. Albro and C. E. Parker, *J. Chromatogr.* **197**, 155–169 (1980).

235. R. L. Harless, E. O. Oswald, M. K. Wilkinson, A. E. Dupuy, Jr., D. D. McDaniel, and H. Tai, *Anal. Chem.* **52**, 1239–1245 (1980).

236. J. N. Huckins, D. L. Stalling, and W. A. Smith, *J. Assoc. Off. Anal. Chem.* **61**, 32–38 (1978).

237. D. L. Stalling, J. Johnson, and J. N. Huckins, in *Environmental Quality and Safety* (F. Coulston and F. Korte, eds.), Vol. 13, p. 12G, Tieme, Stuttgard, 1975.

238. D. L. Stalling, L. M. Smith, and J. D. Petty, ASTM Spec. Techn. Publ. STP-686, 302, 1979.

239. S. Facchetti, A. Fornari, and M. Montagna, *Adv. Mass Spectrom.* **8**, 1405–1410 (1980).

240. R. K. Mitchum, G. F. Moler, and W. A. Korfmacher, *Anal. Chem.* **52**, 2278–2282 (1980).

241. B. M. Hughes, J. R. Troost, J. F. Ryan III, and J. G. Montalvo, Jr., U.S. Environmental Protection Agency, No. 68-01-4887, TAC-NASA.

242. W. C. Brumley, J. A. G. Roach, J. A. Sphon, P. A. Dreifuss, D. Andrejewski, R. A. Niemann, and D. Fireston, *J. Agric. Food Chem.* **29**, 1040–1046 (1981).

243. M. F. Gonnord and F. W. Karasek, *T.S.M.—L'Eau* **77**(5), 221–229 (1982).

244. A. Liberti and D. Brocco, in *Chlorinated Dioxins and Related Compounds: Impact on the Environment* (O. Hutzinger, R. W. Frei, E. Merian, and F. Pocchiari, eds.), Pergamon Series on Environ. Sci., Vol. 5, p. 245–252, Oxford, 1982.

245. J. R. Haas, M. D. Friesen, D. J. Marvan, and C. E. Parker, *Anal. Chem.* **50**, 1474–1479 (1978).

246. W. A. Korfmacher and R. K. Mitchum, *J. HRC & CC* **5**(12), 681–682 (1982).

247. F. I. Onuska, M. E. Comba, and R. Thomson, Canada Centre for Inland Waters Internal Report, Burlington, Ontario, December, 1979.

248. F. I. Onuska, M. E. Comba, and R. Thomson, Canada Centre for Inland Waters Internal Report, Burlington, Ontario, December 1980.

249. F. I. Onuska, M. E. Comba, and R. Thomson, Canada Centre for Inland Waters Internal Report, Burlington, Ontario, 1980.

250. H. R. Buser, *J. Chromatogr.* **107**, 295–310 (1975).
251. H. R. Buser and C. Rappe, *Anal. Chem.* **52**, 2257–2262 (1980).
252. R. L. Harless and R. G. Lewis, Pittsburgh Conference, Atlantic City, 1980.
253. H. R. Buser, C. Rappe, and A. Gara, *Chemosphere* **7**, 439–449 (1978).
254. T. J. Nestrick, L. L. Lamparski, and D. I. Townsend, *Anal. Chem.* **52**, 1865–1874 (1980).
255. G. F. VanNess, J. G. Solch, M. L. Taylor, and T. O. Tiernan, *Chemosphere* **9**, 553–563 (1980).
256. R. Baughman and M. Meselson. An Improved Analysis for 2,3,7,8-TCDD. Presented at 162nd ACS National Meeting, Washington, D.C., 1971.
257. D. W. Kuehl, R. C. Dougherty, Y. Tondeur, D. L. Stalling, L. M. Smith, and C. Rappe, *Environmental Health Chemistry—The Chemistry of Environmental Agents as Potential Human Hazards* (J. D. McKinney, ed.), Ann Arbor Science, Ann Arbor, Michigan, 1980, pp. 245–261.
258. R. M. Smith, P. W. O'Keefe, K. M. Aldous, D. R. Hilker, and J. E. O'Brien, *Environ. Sci. Technol.* **17**, 6–10 (1983).
259. V. Zitko, O. Hutzinger, and P. M. K. Choi, *Environ. Health Prospects* **1**, 47–50 (1972).
260. D. L. Stalling, J. D. Petty, L. M. Smith, and G. R. Dubay in *Environmental Health Chemistry* (J. D. McKinney ed.), Ann Arbor Science, Ann Arbor, Michigan, 1981.
261. J. G. Vos, J. H. van der Maas, H. L. Ten Noever de Brauw, and R. H. Vos, *Food Cosmet. Toxicol.* **8**, 625–629 (1976).
262. J. A. G. Roach and J. H. Pomerantz, *Bull. Environ. Contam. Toxicol.* **12**, 338–342 (1974).
263. C. A. Nilsson and L. Renberg, *J. Chromatogr.* **89**, 325–333 (1974).
264. E. C. Villanueva, R. W. Jennings, V. W. Burse, and R. D. Kimbrough, *J. Agr. Food Chem.* **22**, 916–917 (1974).
265. M. Th. M. Tulp and O. Hutzinger, *Biomed. Mass Spectrom.* **5**, 224–231 (1978).
266. J. D. Petty, L. M. Smith, and J. L. Johnson, American Chemical Society Meeting, Kansas City, Missouri, September 1982.
267. T. Mazer, F. D. Hileman, R. W. Noble, and J. J. Brooks, *Anal. Chem.* **55**, 104–110 (1983).
268. J. D. Petty, L. M. Smith, P. Berggvist, J. L. Johnson, D. L. Stalling, and C. Rappe, American Chemical Society Meeting Kansas City, Missouri, 1982.
269. D. L. Stalling and J. W. Hogan, *Bull. Environ. Contam. Toxicol.* **20**, 35–43 (1978).
270. W. Krijgsman and C. G. van der Kamp, *J. Chromatogr.* **131**, 412–416 (1977).
271. P. Buryan, J. Macak, and U. M. Nabivach, *J. Chromatogr.* **148**, 203–210 (1978).
272. R. F. Brady, Jr., and B. C. Pettitt, *J. Chromatogr.* **93**, 375–381 (1974).
273. N. A. Kirshen, *VIA-Varian Inst. Applications* **16**, 4–5 (1982).
274. L. V. Semencenko and V. T. Kaplin, *Zavodskaya Lab* **33**, 801–804 (1967).
275. A. S. Chau and J. A. Coburn, *J. Assoc. Off. Anal. Chem.* **57**, 389–393 (1974).
276. R. C. C. Wegman and A. W. M. Hofstee, *Water Res.* **13**, 651–657 (1979).
277. J. R. Dahlgran and L. Abrams, *J. HRC & CC* **5**(12), 656–661 (1982).
278. E. Fogelqvist, B. Josefsson, and C. Roos, *J. HRC & CC* **3**(11), 568–574 (1980).
279. M. Deinzer, D. Griffin, T. Miller, and R. Skinner, *Biomed. Mass Spectrom.* 301–304 (1979).
280. H. L. Kopperman, D. W. Kuehl, and G. E. Glass, *Water Chlorination: Environmental Impact and Health Effects*, Vol. 1, (R. L. Jolley, ed.), Ann Arbor Science, Ann Arbor, Michigan, 1976, pp. 327–345.
281. R. H. Pierce, C. R. Brent, H. P. Williams, and S. G. Reeves, *Bull. Environ. Contam. Toxicol.* **18**, 251–257 (1977).
282. D. E. Wells, *Anal. Chim. Acta* **104**, 253–266 (1979).

283. F. I. Onuska and M. E. Comba, in *Recent Advances in Capillary Gas Chromatography*, Vol. 2 (W. Bertsch, W. G. Jennings, and R. E. Kaiser, eds.), Huetig Verlag, Heidelberg, pp. 89–94.

284. K. Lindström, *Das Papier* **31**(12), 517–525 (1977).

285. J. M. H. Bamelmans and M. C. teň Noever de Brauw, *Sci. Total Environ.* **3**, 126–128 (1974).

286. M. A. Ribick, G. R. Dubay, J. D. Petty, D. L. Stalling, and C. J. Schmitt, *Environ. Sci. Technol.* **16**, 310–318 (1982).

287. W. Wideqvist, B. Jansson, L. Reutergardh, G. Sundstrom, and U.-B. Uveno, Statens Naturvardsverk, Report No. 82-01, Wallenbergs Laboratory, Stockholm, Sweden.

288. M. Zell and K. Ballschmiter, *Fresenius Z. Anal. Chem.* **300**, 387–402 (1980).

289. F. I. Onuska and R. D. Thomson, "Toxaphene," Canada Centre for Inland Waters Report to GLWQB, Study No. AMD-80-15, Burlington, Ontario, 1980.

290. T. F. Bidleman and C. E. Olney, *Science* **183**, 516–518 (1974).

291. B. Jansson, V. Reggie, G. Blomkvist, S. Jenssen, and M. Olsson, *Chemosphere*, **4**, 181–190 (1979).

292. D. L. Stalling, J. D. Petty, L. M. Smith, G. R. Dubay, and C. Rappe, 177th National Meeting of the ACS, Honolulu HI, Paper No. 102, April, 1979.

293. M. P. Friocourt, D. Picart, and H. H. Floch, *Biomed. Mass Spectrom.* **7**, 193–200 (1980).

294. M. P. Friocourt, F. Berthou, D. Picart, Y. Dreano, and H. H. Floch, *J. Chromatogr.* **172**, 261–271 (1979).

295. J. C. Peterson and D. H. Freeman, *Int. J. Environ. Anal. Chem.* **12**, 277–291 (1982).

296. D. H. Fine, F. Rufeh, and B. Gunther, *Anal. Letters* **6**, 731–733 (1973).

297. D. H. Fine, D. P. Rounbehler, and N. P. Sen, *J. Agric. Food Chem.* **24**, 980–984 (1976).

298. J. Meili, P. Bronnimann, B. Brechbühler, and J. J. Heiz, *J. HRC & CC* **2**(7), 475–480 (1979).

299. J. K. Foreman and K. Goodhead, *J. Sci. Food Agric.* **26**, 1771–1779 (1975).

300. J. H. Hotchkiss, R. A. Scanlan, and L. M. Libbey, *J. Agric. Food Chem.* **25**, 1183–1189 (1977).

301. H. J. Stan, *Z. Lebensmitt-Untersuch-Forsch* **164**, 153–160 (1974).

302. P. J. Groenen, R. J. G. Jonk, C. van Ingen, and M. C. ten Noever Brauw, *IARC Sci. Publication* **14**, 321 (1976).

303. M. J. Saxby, *J. Assoc. Off. Anal. Chem.* **55**, 9–12 (1972).

304. K. S. Webb, T. A. Gough, A. Carrick, and D. Hazelby, *Anal. Chem.* **51**, 989–992 (1979).

305. T. A. Gough, *Analyst* **103**, 785–806 (1978).

306. R. Farelli, C. Chiabrando, and L. Airoldi, *Anal. Letters* **A11**(10), 845–854 (1978).

307. B. Staucher, F. Tunis, and L. Farretto, *J. Chromatogr.* **131**, 309–316 (1977).

308. E. Stephanou and W. Giger, *Environ. Sci. Technol.* **16**, 800–805 (1982).

309. R. S. Tobin, F. I. Onuska, D. H. J. Anthony, and M. E. Comba, *Ambio* **5**(1), 30–31 (1976).

310. B. G. Luke, *J. Chromatogr.* **84**, 43–49 (1973).

311. R. Deleu and A. Copin, *J. HRC & CC* **3**(6), 299–300 (1980).

312. L. Sucre and W. Jennings, *J. HRC & CC* **3**(9), 452–460 (1980).

313. S. A. Estes, P. C. Uden, M. D. Rausch, and R. M. Barnes, *J. HRC & CC* **3**(9), 471–472 (1980).

314. E. Matisová, J. Krupčik, and O. Liška, *J. Chromatogr.* **173**, 139–146 (1979).

315. G. P. Slater and V. C. Blok, personal communication.

316. G. P. Slater and V. C. Blok, personal communication.

317. R. G. Ackman, J. C. Sipos, and C. Tocher, *J. Fish. Res. Board Can.* **24**, 635–646 (1967).

INDEX